CM:
The Construction Management Process

CM: The Construction Management Process

James J. Adrian, Ph.D., P.E., C.P.A.
Bradley University

RESTON PUBLISHING COMPANY, INC.
A Prentice-Hall Company
Reston, Virginia

Library of Congress Cataloging in Publication Data
Adrian, James J
    CM : the construction management process.

    Includes index.
    1. Construction industry—Management.
I. Title.   II. Title: The construction management process.
HD9715.A2A35   624'.068   80-21624
ISBN 0-8359-0829-1

© 1981 by Reston Publishing Company, Inc.
A *Prentice-Hall Company*
Reston, Virginia 22090

All Rights reserved. No part of this book may be reproduced in any way.
or by any means, without permission in writing from the publisher.

1   3   5   7   9   10   8   6   4   2

Printed in the United States of America

Dedicated to a Changing Construction Industry

Contents

Preface xi

1 **What is CM? (Construction Management)** 1
Construction Management Defined 1
Why CM Evolved 9
Organizational Structure for the CM Process 16
Exercise 1.1 22
Exercise 1.2 22

2 **CM: Advantages and Disadvantages** 23
Benefits of the CM Process 23
Disadvantages of the CM Process 34
When CM is Beneficial 39
Exercise 2.1 41
Exercise 2.2 41

3 **CM Services and Forms of Contracts** 42
Engaging a CM Firm 42
Key Services of the CM Firm 43
Preconstruction Services 45
Construction Services 54
Liabilities of the CM Firm 64
Exercise 3.1 66
Exercise 3.2 66

4 Marketing CM and Selecting a CM Firm   68

Construction Management: A Professional Service   68
Types of CM Projects   69
Marketing the CM Firm   73
Writing a Winning Proposal   77
Skills of the CM Firm   83
Exercise 4.1   85
Exercise 4.2   85

5 Feasibility Estimates and Tax Analysis   86

Importance of the Feasibility Estimate   86
Measuring Project Benefit   87
Owner's Estimated Project Costs   89
Income Tax Related Factors   95
Time Value of Money Concepts   102
Discounted Cash Flow Analysis   109
Office Building Example (No Componentizing)   112
Office Building Example (Using Componentizing)   123
Exercise 5.1   130
Exercise 5.2   130

6 CM Estimates   132

The Importance of CM Estimates   132
CM Estimates: When and Why   133
Unit Cost Estimates   136
Parameter Estimating   139
Factor Estimating   144
Range Estimating   146
Exercise 6.1   150
Exercise 6.2   150

7 Value Engineering   151

Design Inefficiencies   151
Benefits of Value Engineering   152
The Value Engineering Process—The Seven Steps   154
Implementing Value Engineering   163
Exercise 7.1   167
Exercise 7.2   168

Contents

8 **Project Planning and Scheduling** 169
   The Need for Planning and Scheduling 169
   The Bar Chart Model 170
   Network Models and Project Planning 172
   The Critical Path Method (CPM) 174
   CPM Calculations 177
   Time Scale CPM 184
   EST-LST Schedules and Scheduling Resources 187
   Project Scheduling With Limited Resources 192
   Resource Leveling 201
   Exercise 8.1 206
   Exercise 8.2 207

9 **Phasing and Packaging Work** 208
   The Importance of Phasing and Packaging in the CM Process 208
   Phased Construction: The Benefits and Disadvantages 209
   Packaging Work: The Benefits and the Disadvantages 213
   The Role of the Construction Manager in Phasing and Packaging 221
   The Role of Planning Techniques in Phasing and Packaging 223
   Exercise 9.1 226
   Exercise 9.2 226

10 **Controlling Project Time and Cost** 228
   The Importance of Control 228
   The CM Firm's Role in Project Control 229
   Coordination and Communication Meetings 232
   Monitoring Project Performance 235
   Monitoring Contractor Payouts 240
   Correcting Inefficient Construction Operations 244
   The Method Productivity Delay Model 245
   Exercise 10.1 255
   Exercise 10.2 256

   Appendices
   A AIA Standard CM Contract 259
   B AGC Standard CM Contract 269
   C Interest Tables 284
   D Solutions to Exercises 304

   Index 325

Preface

Perhaps the most dramatic event that has changed the construction industry in recent years is the introduction of the CM (Construction Management) process. The CM process is a four party process. Unlike the traditional general contracting process that includes the project owner, designer, and a contractor team, CM is characterized as including a CM firm that serves as an advisor to these three entities throughout the design and construction of a project.

This book devotes itself to CM as a process. This is to be contrasted to the many construction management books that have been written that address management techniques for a contractor. To a degree, some of these techniques are part of the CM process. However construction phase management techniques by themself are not the CM process. CM is a means of delivering a construction project to an owner. There are significant differences between the CM process and general contracting. These differences are set out in the book.

The popularity of the CM process in recent years had led to its accelerated use and misuse. Numerous potential project owners have engaged a CM firm without fully understanding the process. Similarly some firms offer CM services without having the skills to perform the process successfully. Even worse, on occasion, some firms offer CM services to potential project owners without fully understanding the process or CM services themselves.

The lack of understanding of the CM process has led to a fragmented understanding of the CM process. The result has been that not all projects built with the CM process have been a success. This book has the objective of lessening the fragmentation that surrounds the CM process.

The author believes that the CM process described in this book, is the "purest" and most effective CM process. It is characterized by an (1) independent CM firm concept, (2) a broad set of CM services, and (3) an aggressive and creative construction manager. It is these three characteristics that in part differentiate an effective CM process versus the mere use of the CM "buzzword".

The book is essentially two parts. The first four chapters describe the CM process. These chapters contrast the CM process to the general contracting process, the advantages and disadvantages of CM, forms of contracts and CM services, and the marketing and selecting of a CM firm.

Chapters five through ten individually present six of the most important topics and/or management techniques that are part of an effective CM process. Chapters four through seven present preconstruction CM topics, and construction phase topics and management techniques are presented in Chapters eight through ten.

This book should serve the construction, engineering, and business student as well as the practitioner. Each chapter is followed by exercises for the reader to test his or her understanding of the chapter. Suggested or example solutions to each exercise are included in Appendix D. Example CM contracts in common use are presented in Appendices A and B. Interest tables to be utilized in conjunction with chapter 5 are presented in Appendix C.

This book is the outcome of the author's active role in the development of the CM process, and his practice of CM. Over the last few years the author has given over two hundred talks/seminars on the CM process and assisted numerous project owners in their use of the CM process.

The author wishes to thank his family for their patience during the many hours of manuscript prepartion. Most importantly the book would not be possible without the hard work and cooperation of his wife.

J. J. A.

CM:
The Construction Management Process

1

What is CM?
(Construction Management)

CONSTRUCTION MANAGEMENT DEFINED

The term *construction management* (CM) has been one of the most used but least understood terms in the construction industry in the last decade. Some individuals mean by CM nothing more than starting a part of the construction of a project while the design is being completed (referred to as *phased construction*). Others view CM as the implementation of critical-path planning and scheduling for a construction project. Yet other individuals interpret CM to mean the elimination of the general contractor by engaging an individual or company to award separate contracts to several contractors to build component parts of a project.

The many interpretations of CM have lead to much misunderstanding on the part of potential project owners and external parties to the construction process, including lending institutions and surety companies. This misunderstanding has added to the lack of communication and fragmentation that is characteristic of the construction industry.

Construction management is a process. The above-mentioned interpretations of CM are just part of CM. Techniques like phased construction, critical-path scheduling, and packaging of contracts, among others, are tools of the CM firm, not the process itself. These and other CM techniques or management tools are identified in chapter 3 and presented individually in later chapters.

Also not to be confused with the term construction management is better contractor management. Many books have been written with the term construction management in the title, yet they have addressed the management needs of the construction firm rather than the CM process itself. This does not imply that the techniques of good contractor management are separate and distinct from the CM process. Admittedly there is some overlap. However, good contractor management, including the use of management techniques, is not the CM process.

Now that we have indicated what construction management is not, just what is CM? Although not necessarily a workable definition, the following might be set out as an acceptable definition of the CM process.

CM (Construction Management) is a process by which a potential project owner engages an *agent*, referred to as the CM, or Construction Manager, to *coordinate and communicate* the *entire project process*, including project feasibility, design, planning, letting, construction, and project implementation, with the objective of *minimizing the project time and cost, and maintaining the project quality.*

The definition has four important terms that have been set out in italics. These terms are what make CM different from general contracting, turnkey construction (the process by which a single contractor bids the entire project and then proceeds to award subcontracts for all the construction), and design-build. These four distinguishing characteristics are illustrated in Figure 1.1.

Agency

The term *agent* in the definition of CM results in the construction management (CM) firm having a different legal relationship with the project owner than the traditional general contractor. The general contractor is legally an "independent contractor." As an independent contractor, the general contractor does not have the authority to act for the project owner in dealing with external parties. The independent contractor acts in its own behalf in all of its business dealings.

Unlike the independent contractor, an agent (the CM) has an obligation to serve the owner as if he or she were an employee of the owner. An agent has legal authority to represent the owner and to carry out business dealings in the owner's behalf. Legally, an agent has the following contractual obligations to its employer (or owner in the case of the CM firm).

1. Loyalty
2. Obedience and performance
3. Reasonable care
4. Accounting
5. Information

Because of its agency relationship, the CM firm has a right to select contractors for the owner, to purchase materials for the owner, and to "manage" individual prime contractors in the owner's behalf. For example, if the contractor has not performed work according to specifications, the CM firm acting in the owner's behalf, can withhold payment from the contractor. In acting in the owner's behalf, the CM firm has the same legal liability for its actions or decisions as the owner would have. For example, if the CM makes

Construction Management Defined

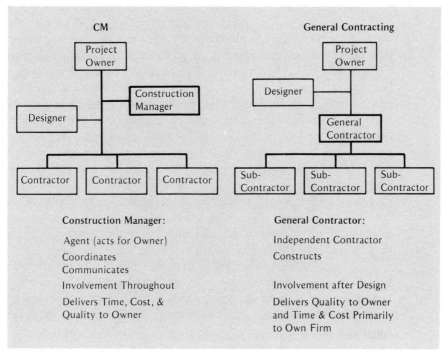

Figure 1:1 Construction management versus general contracting

a decision that leads to one of the on-site contractor's being delayed in its work schedule and subsequently to additional contractor costs, the CM firm may be held responsible for the damages the contractor incurs. It should be noted that the owner is also indirectly liable for these damages and other claims that arise due to the actions or decisions of the CM firm.

As noted in the above list of agency/employer (or CM firm/project owner) requirements, the agent is expected to demonstrate loyalty to its employer. This is a very important requirement. It means that if the potential construction project owner can show that the CM firm has not been loyal in its fulfillment of its contract duties, the owner can correctly claim that the CM has breached its contract with the owner. The question is relevant to the discussion of whether the CM firm should give the project owner a guaranteed maximum contract for the proposed construction project or whether the CM firm should perform some of the construction work (for example, provide the labor and material for doing the concrete and masonry work) with its own resources. The advantages and disadvantages of both the guaranteed maximum and the CM firm's doing some of the construction work raises the question of whether the CM firm can remain independent and loyal to the owner given either of these two conditions.

Assume, based on an initial review of the scope of a proposed project, the CM firm is asked to give the project owner a guaranteed maximum dollar amount for the project, including the CM firm's fee. Let us assume that the CM firm complies and agrees to a guaranteed maximum dollar amount of $2 million. The CM firm proceeds to assist the owner in awarding construction contracts and in supervising the construction process. However, due to unexpected bad weather and some material shortages, the project is delayed. A revised estimate indicates that the project will cost in excess of the $2 million. Given the existence of the guaranteed maximum dollar contract, it can be argued that the cost overrun may result in the CM firm having to take a self-interest in the project in order not to jeopardize its equity and financial well-being. If this is the case, it may be difficult for the CM firm to maintain its loyalty to the project owner. In other words, the existence of the guaranteed maximum dollar CM contract may in part result in a CM breach of contract because it may force the CM firm to act in its own self-interest rather than in behalf of the project owner.

Some individuals also argue that the loyalty requirement of the CM firm/owner agency relationship is jeopardized by the CM firm being allowed to perform a portion of the construction project work with its own resources. Under its CM contract with the owner, the CM firm is an agent. In performing construction work on the project, the same firm is an independent contractor. The question arises as to whether the CM firm, under its CM contract, can independently award the construction work to itself and subsequently objectively manage and supervise its performance of the construction work.

Coordinate and Communicate

The second important term of the definition of CM was the fact that the CM firm *coordinated* and *communicated* the entire project process. Perhaps more important than the fact that the CM firm is to coordinate and communicate is the fact that the definition does not state that the CM firm neither designs nor builds the project.

Although the CM firm assists in the design phase of the project, it does not design it. The design remains the responsibility of the designer, the architect and/or engineer. During the design phase, the CM firm serves as an adviser to the designer and project owner regarding the economical choice of materials and methods, and the availability of materials, labor, and contractors for the proposed design. The integrity and structural stability of the project design remains the responsibility of the design team.

The definition of CM also fails to list actual construction of the project as a responsibility or duty of the CM firm. Coordinate and communicate does not infer construct. Despite the inference that the CM firm does not

Construction Management Defined

construct the project with its own resources, many CM firms do in fact do varying amounts of construction work under the umbrella of its CM contract.

Some CM firms justify doing a minor portion of the construction work with their own forces based on the premise that parts of the work are difficult or impossible to put out for bid to external contractors. Other CM firms justify awarding themselves work when they can perform the work in question for less than the bids received from external contractors. Still others, upon giving a guaranteed maximum dollar contract for the entire project to the owner, hold open the possibility of doing some of the construction work as a means of ensuring the owner a total project cost less than or equal to the guaranteed maximum amount. Should external contractors bid amounts exceeding the component parts of the budgeted guaranteed maximum, the CM takes on the responsibility of performing the component part or parts for the amounts budgeted.

Although each of the noted reasons the CM firm might give for performing some construction work has merit, the fact remains that it is not truly CM. What it is, is a mixture of the CM process and contracting. One might even argue that the mixture of the two processes results in a process that shouldn't be referred to as CM. However, in practice, this combination of performing CM services (as discussed in chapter 4) and doing some construction work is marketed as CM.

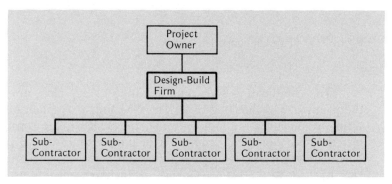

Figure 1:2 Design-build construction process

Some individuals confuse CM with the concept *of design-build*. Actually, CM and design-build are opposites. In the design-build process, a single firm provides the potential project owner both project design and construction services. However, in a true CM process, the CM firm provides neither the design itself, nor the construction. The design-build process is illustrated in Figure 1.2.

The CM process also differs from turnkey contracting. In turnkey contracting, a firm takes responsibility for the construction phase (both time and cost), but in fact subcontracts *all* of the work. The subcontractors have a contract with the turnkey contractor. In the CM process, the contractors have prime contracts with the owner. Usually there is no contract between the CM firm and the project contractor. (An exception to this "no-contract" relationship between the CM firm and the contractor is discussed in a following section of this chapter.) Another difference between CM and turnkey construction is that the CM firm is usually more active in preconstruction services than the turnkey contractor. The turnkey construction process is illustrated in Figure 1.3.

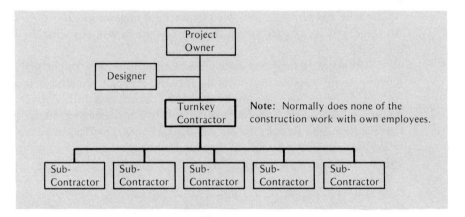

Figure 1:3 Turnkey construction process

The coordinating and communicating of a project can be viewed as the planning and directing of it. The CM firm is "captain of the team," a team that includes the project owner, the designer, the construction team, material and equipment vendors, labor unions, and external parties such as bonding companies, lending institutions, and governmental regulatory agencies. Each of these "team members" is an important link in the success or failure of the construction project.

In effect, the CM firm has the responsibility of leading all of these team members, each with its own self-interests and objectives, toward a common objective. One should not discount the CM firm's difficulty in accomplishing this objective. The construction process has been plagued by adverse relationships among the parties. For example, the design professional has not always seen eye to eye with the construction contractor and vice versa.

Although one can list specific duties and management techniques that should be part of the CM services (these will be discussed in later chapters), one should not discount the CM firm's ability to get all of the team members heading toward a common goal. Perhaps this is the most important service the potential project owner should expect of the CM firm. The CM firm should be promoting consultation rather than confrontation!

Involvement in the Entire Project Process

The third, and perhaps most distinguishing characteristic of the CM process is the fact that a single firm, the CM firm, is involved in the entire project process, including project feasibility, design, planning, letting, construction, and implementation. The CM firm's involvement in each of these phases is dependent on the complexity and time frame of a project. Notwithstanding that the CM firm's involvement in specific project phases will vary, the fact remains that the CM firm is involved in more than one phase of the project.

The non-CM approach to a project is characterized by individuals limited to involvement in a single phase of the total project. A designer's input, for example, is essentially limited to preconstruction. Although it is true that the designer may contract to inspect the project and approve contractor pay requests, this role is more of a policing role than active involvement. Similarly, in the non-CM process, the contractor role focuses on the construction phase of the project. More often than not the contractor has no input to the feasibility or design phases, and perhaps even the planning phase of the project.

One can correctly argue that it is difficult if not impossible to optimize time and cost for a project unless there is a single firm active and responsible throughout the entire process. The CM process is consistent with the *systems approach* to management. The systems approach indicates that in order to optimize the solution to a problem, one must first interrelate all of the component parts. It may follow that the component parts may have to be performed suboptimally. Looked at another way, it doesn't follow that optimal solutions for component parts can be summed to yield the optimal total solution.

In regard to the CM process and the systems approach to management, the CM firm is involved in the entire project process, including feasibility, design, planning, letting, construction, and implementation. The CM's total involvement enables it to interrelate all of the component parts of the project process. It can be argued that without a single firm's involvement throughout the entire process, and the recognition of the interrelationships of project feasibility, design, planning, letting, construction, and implementation, it is difficult if not impossible to minimize the construction project time and cost and maintain the project quality. For example, it is true that the designer of a project must take into account the cost and availability of the components of the project design if the design is to be of the highest value to the project owner. The CM firm must bridge the gap between design and construction to ensure an optimal design, as it also must interrelate other phases of an owner's project.

Minimize Project Time and Cost and Maintain Quality

There are three variables that are relevant to the success or failure of almost every construction project: time, cost, and quality. These three variables and the potential project owner's concern with each is illustrated in Figure 1.4. The CM process has the objective of controlling each of these three variables.

Worth of Project	Time	Cost	Quality
Ability to Initiate Project		X	
Ability to Finance Project		X	
Project's Rate of Return	X	X	
Potential Legal Problems	X	X	X
"True" Cost of Project	X	X	X
Utilization of Project	X		X

Figure 1:4 Three variables of a proposed construction project

The time variable is important to an owner because the profitability of the proposed project is dependent on its completion date. This is especially true for investment property. For example, the profitability of an apartment or office building is very sensitive to the date at which the rental revenue can be generated.

The completion time of a project is also important because project delays usually result in additional project costs. Usually time is money in regard to a construction project. This is especially true given the high interest costs associated with financing a project during construction. Few potential project owners have unlimited funds. Even those owners with sufficient funds for a project would like to maximize the size or quality of their project for the dollar budget available. Therefore, cost is almost always of major concern. This is true regardless of whether the project is funded with private or public money.

Although project time and cost are of utmost concern to the potential project owner, many owners have been unwillingly subjected to excessive time and cost overruns. Reasons for these overruns vary from uncontrollable events such as unusual weather, to poor management practices of contractors. Regardless of the cause, the fact remains that the owner may not always obtain the time and cost budgeted for in the project.

An objective of the CM process is to deliver the owner less project time and cost than alternative systems. Although coordination and communication are the services or skills of the CM firm, the measure of the effectiveness of these services or skills is the project time and cost.

Quality, although an important construction project variable, is not as

Why CM Evolved

subject to variation as project time and cost. The project contract documents, especially the project specifications, play a major role in setting out and controlling the quality of the project. However, the CM process attempts to maximize the quality of the proposed project during the design process, and it provides the control procedures during the construction phase of the project to ensure that quality is achieved. The CM process does not sacrifice project quality at the expense of minimizing project time and cost. The objective is to minimize the project time and cost and maximize or at least maintain the project quality.

WHY CM EVOLVED

The use of CM for building projects has increased rapidly over recent years. More potential owners, especially in the private sector, have turned to hiring a CM firm to manage the building process. With this increase in demand comes an increase in the number of firms offering CM services. If one merely thumbs through a community's Yellow Pages for the last five years, one can observe the increasing number of firms offering CM services.

The first use of the CM concept is difficult to pinpoint. Some firms, including developers and some owners of projects using in-house expertise, claim to have performed CM services many years, but not titling the services as such. These claims can be disputed. Nonetheless, it is true that several progressive firms have been performing some if not all of the services of a modern-day CM firm for many years.

Although the first use of the term CM (as defined in the prior section) can be disputed, the fact remains that the term came into widespread use in the late 1960s. From that time the understanding of the concept of CM and the use of the process has increased steadily. Given the rather recent emergence of the conception and use of CM, one might ask the question, "Why did CM evolve?" Did the need create the concept, or did the concept create the need? A summary of the reasons often given for the emergence and growth of the CM process is given in Figure 1.5. The reasons are listed in order of importance as judged by the author; the first reason listed being the most important reason, etc.

CM Growth
1. Failure of traditional method to attain owner's time, cost and quality objectives
2. Limited liability and bonding requirements of CM firm
3. Compatibility of CM process with increased project complexity
4. Stimulation of the use of the CM process by public agencies

---

Figure 1:5   Reasons for growth of the CM process

Failure to Control Time and Cost

As noted in Figure 1.5, the CM concept undoubtedly emerged in great part due to the failure of the traditional three-party process (as described in this chapter) to deliver two of the three important construction project variables: time and cost. This is not to say that the traditional process always failed. On some projects the three-party process, consisting of the owner, the architect or engineer (referred to as the A/E serving as the project designer), and contractor team, has proved an effective and efficient means of building a project. However, numerous examples of projects that have incurred time and cost overruns relative to the owner's initial estimates have been documented.

Examples of such time and cost overruns could be cited in every community in almost every part of the country. Admittedly not all of the overruns have been extreme, nonetheless they have resulted in the project owner's dissatisfaction with the building process and have on occasion led to financial ruin of the contracting firm, the project owner, or both.

The causes for the inability of the traditional building process to deliver time and cost often stem from the following.

1. The inability of the A/E team to give reliable budget estimates during and after the design phase.
2. The absence of construction expertise in the design phase.
3. The inability of the contractor team to implement good management practices during the construction phase of a project.
4. The linear characteristic of the feasibility, design, letting, and construction phases of the traditional three-party building process.

The design team has often been justly criticized for not being good estimators of construction costs. Some would suggest that the A/E's estimates are at best "guess-timates" with a range of accuracy of plus or minus 40 percent of the actual cost of construction. This poor accuracy leads to a less than ideal design and often relates to later delays in letting of contracts and contract disputes.

In defense of the design team, it should be noted that preliminary and design estimating services have been forced upon the A/E. The A/E often doesn't feel comfortable giving these estimates, as estimating is not typically a significant part of their formal education. Secondly, the typical A/E's lack of involvement in the actual construction results in little on-site knowledge of up-to-date field costs. Without this more direct involvement in on-site observation and documentation of construction costs it is somewhat unrealistic to demand accurate estimates from the A/E firm. CM is held out as a partial solution to this inability of the A/E to give early and accurate estimates. As discussed in chapters 5 and 6, budgeting and estimating are important CM services.

Also relevant to potential inefficiencies in the design phase is the lack of sufficient construction expertise. Too often the total design of a construction project is dependent on the value judgments of one individual or firm, i.e., the A/E design firm. This is not to say that the A/E firm is lackadaisical in its duties. Instead, we are saying that any one individual or firm may have rather fixed ideas and values.

Additionally, the failure to have an accurate conception of construction costs or construction methods may lead to ill-fated or nonoptimal choices of materials or methods. For example, a designer's misconception about the productivity of a mason and the resulting unit cost of a concrete-block wall for an interior partition may lead to the specification of this type of construction instead of a poured concrete wall. If this conception is wrong, the design results in an added cost to the owner. Another example of an excessive cost design results when the A/E firm designs beam-and-column connections that, while structurally sound, may be excessively expensive to build.

Attention is often drawn to the expensive and time-consuming delays that occur in the field production process. Poor crew balances, craftsmen waiting for materials, and equipment breakdowns are validly cited by critics of the building process as causes of delays. Less frequently, critics point to the lack of cost-effective designs. This does not imply that the A/E always produces good value, cost-conscious designs. More likely, the critics' unfamiliarity with the design process is the reason for frequently looking to the contractor team as the culprit responsible for project time and cost overruns.

Because they are more obvious, project method productivity delays in the field can often be corrected and controlled with proper management techniques. However, once the design team produces improper drawings and specifications, these inefficiencies are "ground into the project forever." No subsequent management practices, even good on-site construction management, can remove the inefficiencies.

Consider the case of the design of a kitchen sink. European construction of this element is much more dependent on "wet construction" (i.e., concrete and mortar), than the "dry construction" (i.e., carpentry) that typifies the U.S. design. This is due to the availability of the materials in the two areas. The important point is that the cost of the kitchen sink to the owner of the project is in great part dictated by its design. The European system involves a different material cost. Perhaps more important is the construction labor costs that result from the two designs. The European system implies the use of more labor crafts and the need to bring back several labor crafts more than once. This leads to a larger labor cost for the installation of the sink. Independent of the productivity of the craftsmen, much of the labor cost is predetermined at the design phase.

Advocates of the CM concept hold out that expensive and wasteful design inefficiencies can be reduced in part by introducing more construction expertise into the design phase of a construction project. This is done by having the CM firm serve as an adviser and/or critical reviewer of the A/E's design as it evolves.

One of the more common reasons cited for project time and cost overruns is the poor management practices of the construction firm or firms building the project. Included in the specifics often cited are the contractor's lack of adequate project planning and scheduling of labor, material, equipment, and cash. The lack of adequate cost accounting reports leading to a lack of necessary control has undoubtedly led to project time and cost overruns.

As is true of the design profession, not all construction firms have inadequate project management practices. However, the fact remains that the operation of a construction firm has proved to be a high-risk business. Construction contracting has one of the highest bankruptcy rates of all businesses. High on the list of causes for these financial failures are the contractor's lack of project planning, scheduling, and controlling.

It is not unusual to find that in general, contractors have exhibited less than adequate project management practices. Most construction firms have been started by individuals who came "up through the crafts." In other words, the founders of most construction firms have a production background. Few have had experience in planning, scheduling, and controlling work. Given his interest in production, the craftsman often has a distaste for forms and paperwork. Unfortunately, these forms and "paperwork" are a necessary ingredient to the control of project time and cost.

Contractors have been known to operate by the "seat of their pants." This approach to project management has also led to time and cost overruns, and has led the potential project owner to look for alternative means of obtaining the needed project; one of these means being the CM process.

Even if one assumes that the design and construction teams for a proposed project perform their duties optimally, the traditional means of building a project may lead to time and cost inefficiencies by the very nature of the process. There are many phases a construction project goes through from its conception to its completion. These individual phases can be grouped into one of the following overall project phases: Design phase, letting phase, and construction phase.

More often than not, the completion of these phases is a linear process. Project contracts are let after all of the project drawings and specifications are complete. In turn, the construction phase is started after the construction contract is awarded to a general contractor who normally awards part of the work to subcontractors. Because all of the construction work is awarded to a single contractor, referred to as the general contractor, no work can be initiated until all the project drawings are complete.

Why CM Evolved

The linearity of the traditional building process is illustrated in Figure 1.6(a). Many would suggest that the very nature of this process leads to an unnecessarily long time period for building a project. For example, it is common for some public construction projects to take an equal time period to design, let, and construct. A project that takes one year to build, may take one year to design, and one year to let contracts. A one-year construction project in effect becomes a three-year project. With inflation, the cost of construction can increase substantially between the time of inception of the project to the time construction actually begins. In fact, this increase in costs between inception and actual start of construction has on occasion resulted in a project not being built.

As illustrated in Figure 1.6(b), the CM process offers an alternative to the linear building process. Specifically, the concept of *phased construction* or *fast-track construction* has led to a merging of the design, letting, and construction phases of the building process. Phased or fast-track construction is just one of many management techniques at the disposal of the CM firm. The details of using phased or fast-track construction are the subject of chapter 9. As illustrated in Figure 1.6(b), it becomes possible to shorten the duration of the construction process by merging individual phases of the total process.

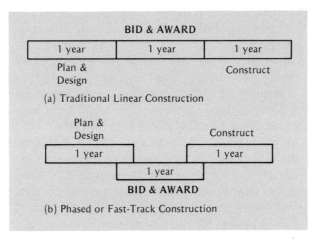

Figure 1:6 Traditional process versus phased construction

Limited Liability and Bonding Requirements

In the previous section, the creation and wide usage of the CM process were identified as related to the failure of the traditional building process to provide adequate control of project time and cost. One might say that the need to control time and cost resulted in the creation of the CM concept. It is also true that the evolution of CM has come about because of aggressive marketing by the individuals offering CM services; a concept was created and then a market was established for its use.

The point to be made is that several firms have begun offering CM services for reasons beyond the fact that the process works or is demanded by potential project owners. Specifically, several firms, especially general contracting firms, have seen the offering of CM services as a means of reducing or eliminating the liability associated with general contracting and also the need to be dependent on the ability to secure bonding, which is also normally part of being a contractor.

Construction contracting has been historically a very high-risk venture. Year in and year out, construction contracting has one of the highest financial failure rates of any industry. Many of these failures can be traced to the competitiveness of the lump sum "hard dollar" contracting process. In this process a number of contractors submit project bids in sealed envelopes with the contract usually being awarded to the lowest bidder. This process has a tendency to "drive-down" the amount of profit a contractor can put in a bid and still remain competitive. Equally important is the fact that the bid submitted by a contractor becomes the contract price. This is usually what the contractor gets paid regardless of the actual construction costs. Given all the uncertainties that result in the building process, the contractor's actual building costs frequently exceed the bid cost. This is the risk of construction contracting.

Although CM services are offered by individuals or firms of various backgrounds, including architect/engineer firms and management consulting firms, a majority of CM firms evolve from a construction contracting background. Some of these contractor-turned-CM firms have made the switch because of the limited liability of the CM firm relative to the construction firm.

Although actual CM contracts vary (as is discussed in chapter 3), initially the CM process and contract were thought of as a professional service and contract. The contract signed between the CM firm and the project owner is much the same as the designer-project owner contract in regard to liability of the nonowner party. In this type of CM contract, the CM firm manages the building process for the owner from start to completion for a fixed fee. The CM firm makes no legal commitment as to the total project cost. While the CM firm would make project cost estimates, for example, $2 million, no cost guarantees would be given. If the project costs were actually $2,200,000, the unexpected $200,000 of project cost would be the liability of the project owner and not the CM firm.

The above type of CM contract remains in common use today. Another common type of CM contract discussed in chapter 3, is, the guaranteed maximum CM contract. The benefits and disadvantages of these two different forms of contract are also discussed in chapter 3.

Some general contracting firms have entered the CM process to lessen their liability as construction contractors. This is not to imply that these firms are not good CM firms; only that there has been an incentive to become a CM firm given the nonguaranteed maximum CM contract.

The incentive to be a CM firm is also created by the role bonding plays in the construction process. More often than not the CM firm does not have to provide a performance bond for the construction process. For the contracting-turned-CM firm, this is a considerable advantage. The ability to obtain a performance bond often dictates the very ability of the general contractor to carry out work. The bond insures the owner that should the construction firm or one of its subcontractors fail to perform its contract obligations, the surety company writing the bond will step in and see to it that the project is completed. More often than not, the general contractor provides the bond for all of its subcontractors.

The ability of a contractor to get a bond is dependent on a firm's financial strength. Almost always the owners of the construction company have to pledge their personal assets, including their house and other personal wealth, in order to get a bond. This is another unfavorable aspect of being a general contractor.

The CM usually avoids the need to be bonded because each of the separate contractors to which work is awarded are required to provide a bond. Bonding and the CM process are further discussed in chapter 2.

CM as a Solution to Increased Project Complexity

Recently construction projects have increased in size and complexity. The increase in complexity can be witnessed in residential, building, and heavy and highway construction. Perhaps the greatest increase in the complexity of the construction project is in building construction.

The subsystems that make up a project, the electrical system, heating and ventilating system, piping system, structural system, etc. have all increased in complexity. For example, recent changes in energy systems have had a significant impact on the fabrication and installation of electrical and heating and ventilating systems. As each of a project's subsystems increase in complexity, the integration of the subsystems into a total "workable" project becomes more complicated as to managing the time, cost, and quality of construction.

Some individuals see the CM firm as the entity responsible for understanding and managing the entire building process. As projects become more complex, there is more and more need for a single entity to serve as overseer for the entire process. This single source is the CM firm. Too often the increased complexity is beyond the comprehension of the general contractor. This is in part the result of the general contractor's emphasis on the areas in which it directly performs work, for example, concrete, carpentry, and perhaps excavation and masonry work. In the past it has not been the responsibility of the general contractor to understand the technical aspect of the work performed by its subcontractors.

This lack of knowledge of the technical aspect of its subcontractor's work is a cause of the inability of the general contractor to efficiently plan and control project time and cost. The problem is compounded by the increasing sophistication and complexity of projects. The CM is being called upon to possess a broad range of knowledge of the technical aspects of all phases of construction and a broad range of management "know how." In effect, the increased complexity of construction projects has enhanced the need for the CM firm and CM process.

Public Agencies' Stimulation of the CM Process

Public construction makes up a significant portion of total annual construction expenditures in almost every country. In the United States approximately 25 to 35 percent of all annual construction expenditures are for public construction.

In the late 1960s and early 1970s, the CM process was successfully marketed to several large public agencies, such as the General Services Administration (GSA) and several branches of the military. These and other federal and state agencies quickly became major users of CM. Some of these agencies used the CM process almost exclusively to build their total-budgeted construction projects.

The widespread use of the CM process by public agencies assisted in increased awareness of the CM process and its workings in the private sector. Manufacturers, schools, and eventually speculative projects started to use the CM process.

It is interesting to note that although the initial use of CM by public agencies played a large role in making the process popular, today the use of CM is more common in the private sector. This is due in part to some of the constraints that have evolved with the use of the CM process by public agencies. For example, laws prohibit some public agencies from starting a project without knowing the total cost of construction. This often eliminates the use of phased or fast-track construction, one of the more common practices at the disposal of the CM firm. Nonetheless, in the early years of CM, public agencies had much to do with the development of the process.

ORGANIZATIONAL STRUCTURE FOR THE CM PROCESS

The process by which construction projects have historically been designed and built has included three entities: the potential project owner, a design team consisting of an architect, engineer, or both, and a construction team.

Organizational Structure for the CM Process

The construction team has usually been led by a single firm, often referred to as the general contractor. The general contractor in turn usually subcontracts a significant portion of its work.

It is true that other parties in addition to the owner, designer, and contractor team play a role in the evolution of a construction project. For example, a lending institution lends financial support to the owner and the contractor, and a surety company provides bonds to the contractor team. However, these external parties play little or no role in the actual design or construction phases of a project. Therefore, it is common to refer to the traditional means by which the construction project has been built as a three-party process. The organizational structure by which the three-party process brings the project owner, designer, and contractor team together is shown in Figure 1.7.

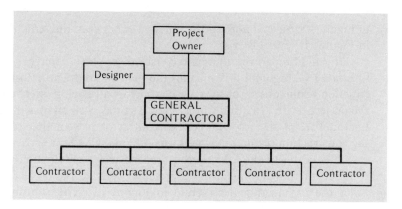

Figure 1:7 The general contracting process

On certain types of construction projects two of the entities shown in Figure 1.7 may be represented by the same firm. For example, in residential construction, a contractor may provide the owner a packaged design (in effect providing design-build services). Or in heavy and highway construction, a public agency may represent the owner and also provide the design. Nonetheless, the fact remains that there is still an owner, a design function, and a construction function.

The CM process is a four-party process. In addition to the potential project owner, the designer, and the contractor team is the CM firm. In the truest form of CM (as defined in the initial section of this chapter), the CM firm is the manager while the architect/engineer is the designer, and the construction firm is the builder.

The most common organizational structure for the CM process, and the structure that is most consistent with the definition of CM, is illustrated

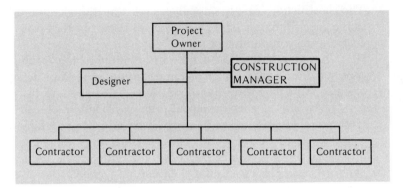

Figure 1:8 The CM process

in Figure 1.8. Several noteworthy differences between the structures illustrated in Figures 1.7 and 1.8 exist.

In addition to the four-party versus three-party difference, the structure illustrated in Figure 1.8 is missing the general contractor. Instead, the construction team consists of separate contractors all having direct contracts with the project owner. In other words, the CM process eliminates the general contractor. Part of the role the general contractor played, that of coordinating the construction team, is now performed by the CM firm.

Advocates of the CM process argue that the elimination of the general contractor removes some of the contractor overhead the project absorbs when using the organizational structure in Figure 1.7. The subcontractor marks up its work a certain percentage for its overhead and profit, say 20 percent, and the general contractor in turn marks up the subcontractors' total bid cost an additional amount. In other words, there is a doubling of contractor overhead and profit. Because there is a flattening out of the construction firm organizational structure (a horizontal structure rather than a vertical), there is less potential for this doubling of contractor overhead and profit in the CM process.

The critic of the CM process might counter the lessening of double overhead and profit by arguing that the awarding of separate contracts creates more project overhead. This is based on the argument that each contractor awarded a contract, including specialty contractors, must provide supervision that was previously performed by the general contractor. The CM critic also argues that as the number of contractor packages increases (there are usually more contractors on the construction team in the CM process than in the process shown in Figure 1.7), additional contractor overhead is incurred because each contractor has an initial move-in and set-up cost and the cost of supervising and managing the construction team increases as the number of separate contractors increases.

Independent of the arguments for or against the elimination of the general contractor, it should be noted that in the CM organizational structure each engaged contractor has a direct contract with the owner. In effect, each contractor is a *prime* contractor. This process differs from that shown in Figure 1.7, where much of the construction work on a project is performed by subcontractors. These subcontractors have a contract with the general contractor, who in turn has a contract with the project owner.

It should be noted that in the CM process shown in Figure 1.8, the CM firm does not have privity of contract with the individual contractors. Owner-requested bonds are the responsibility of the individual contractors. This differs from the process shown in Figure 1.7, in which the general contractor may provide the bond or bonds for the entire construction team.

The organizational structure shown in Figure 1.8 implies that there is no legal contract relationship between the designer and the CM firm. Each has separate contracts with the project owner. In effect, the designer is not working for the CM firm nor is the CM firm working for the designer.

There is a question as to whether the potential project owner should first engage the CM firm or the design firm. The higher level of the horizontal line connecting the CM firm to the organizational chart in Figure 1.8 implies that the CM firm is the first firm engaged. In practice, the owner may engage the design firm first. However, there are two major disadvantages to engaging the design firm before the CM firm. For one, many of a CM firm's services are preconstruction services. If the CM firm is engaged after the design firm, there is a distinct possibility that the CM firm may not have an opportunity to exercise all of its preconstruction services. For example, part of the design may be established before the CM firm's input; this reduces the potential benefits of the CM process because the CM firm will be unable to contribute its input regarding the cost and availability of materials and labor that pertain to the chosen design.

Another disadvantage to engaging a design firm before the CM firm is selected is that the CM firm's inability to play a role in selecting the design firm may limit the potential contribution of the CM firm to preconstruction phases of a project. One of the constraints on effective use of the CM process is for the owner to engage a design firm that neither appreciates the potential contributions of the CM firm nor is willing to cooperate with the CM firm's practices or decisions. If the CM firm has a role in the selection of the design firm, the chances of a noncooperative design firm being selected diminishes. In effect, there is a better chance that a team process, each member of the team appreciating the attributes of the other member, will evolve if the CM firm is the initial firm engaged by the project owner.

The potential weakness of the CM organizational structure is that there is no one member of the construction team legally responsible for the overall construction phase of the project. It can be argued that without a single construction firm being in charge or legally responsible, the construction phase of the project will lack continuity and disputes and disruptions in the work process may evolve. Although the CM firm serves as "captain of the team" for the project, it can be argued that the CM firm's lack of contract relationship with the individual contractors does not give the CM firm authority over the contractors and therefore technically there is no entity "in charge." This is a debatable point in that it can be argued that one does not need a contract relationship to have authority. This "lack of authority" issue is discussed in more detail in chapter 2.

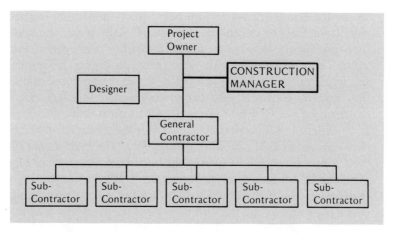

Figure 1:9 CM process that includes general contractor

To offset this potential problem, another CM organizational structure is sometimes used. It is shown in Figure 1.9. This organizational structure includes a general contractor as well as a CM firm. Although not in use as often as the structure illustrated in Figure 1.8, the structure in Figure 1.9 can be used to answer the lack of authority argument. In this organizational structure, the general contractor is legally responsible for the performance of the entire construction team. The specialty contractors are subcontractors as they were in the non-CM organizational structure shown in Figure 1.7.

The CM structure illustrated in Figure 1.9 may offset the lack of authority issue, but it has several disadvantages relative to the structure in Figure 1.8. For one, there is the potential doubling of general contractor/subcontractor overhead and profit. Secondly, the existence of both the general contractor and CM firm leads to a duplication of services, and perhaps even worse, the possibility that some owner-needed services may not be performed because the general contractor looks to the CM firm and vice-versa, with the result that neither performs the service expected. It is also true that the engagement of a single contractor responsible for the entire construction process

Organizational Structure for the CM Process   21

may make it difficult for the owner or CM firm to phase or fast-track the construction work.

Another possible CM structure is illustrated in Figure 1.10. In this structure each of the separate contractors has a direct contract with the CM firm. This results in the CM firm having legal authority over the contractors. However, this structure puts the CM firm more in the role of an *independent contractor* than an owner's agent. The result is that the CM firm is more of a general contractor or turnkey contractor than a firm providing CM services.

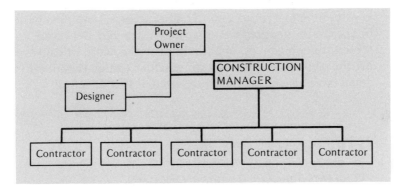

Figure 1:10   CM process with CM firm/contractor contract relationship

In practice, the term CM or CM firm is used to characterize other organizational structures. For example, an owner may engage a construction firm on an hourly basis to work with the design firm to ensure "good construction value" to the project drawings and specifications. The engaged construction firm then may proceed to perform some or all of the construction work. This process has also been referred to as CM. This and the many other deviations in organizational structure among the owner, designer, and contractor are not really the CM process if one holds to the previously given definition.

The use of the term CM to refer to many different structures only further fragments an already fractured construction industry vocabulary. Using the same term to represent different structures and processes confuses an owner and enhances the possibility that the owner will obtain services that yield fewer benefits than anticipated. This is not to say that an organizational structure like that shown in Figure 1.10 cannot be effective for delivering time, cost, and quality to an owner. However, these structures should not be confused with the CM process or structure illustrated in Figure 1.8.

EXERCISE 1.1

INEFFICIENT PROJECTS

This chapter discussed the fact that the major reason for the evolution and use of the CM process relates to the failure of non-CM processes to deliver minimum project time and cost along with acceptable quality of construction to a project owner. List projects, both on a national level and in your own geographic area, that were not delivered consistent with these three objectives. In addition, cite the cause or causes for not attaining the objectives and indicate whether CM as described in this chapter could have helped prevent the problems that occurred.

EXERCISE 1.2

PROBLEMS OF PRIVITY OF CONTRACT

This chapter discussed several different organizational structures for the CM process. Specifically, the organizational structure shown in Figure 1.8 was identified as the most common and preferred CM structure. However, the structure illustrated in Figure 1.10, which included the individual contractors signing agreements directly with the construction manager, was suggested as partly offsetting the potential "responsibility without authority" problem sometimes associated with the CM process. List and discuss disadvantages of the CM firm having direct contracts with the contractors. You should cite disadvantages to the project owner and contractors as well as the CM firm itself.

2

CM: Advantages and Disadvantages

BENEFITS OF THE CM PROCESS

The CM process has been held out as yielding various benefits to project owners. Although it is true that the project owner has on occasion received fewer benefits than anticipated, the fact remains that the growth of the CM process is evidence that CM has yielded benefits that exceed the cost of the CM service.

As noted in the definition given in chapter 1, the CM process has the objective of delivering the project owner minimal project time and cost and acceptable, if not optimal, construction quality. As such, it follows that these three project variables—project time, cost, and quality—are the benefits of the CM process. Given the objective of minimizing project time and cost and maintaining or maximizing quality, the question arises as to what specific benefits the CM process yields the project owner in achieving these objectives.

Benefits of the CM process fall into one of three categories:

1. Project owner benefits resulting from the CM organizational structure.
2. Project owner benefits from the services of the CM firm.
3. Benefits to nonowner entities, including the project design and contractor teams.

The project owner benefits are to be emphasized, because without these potential project owners would not use the process. Any nonowner benefits can be viewed as "fringe" benefits in that they promote the willingness of nonowners to offer or cooperate with the CM process.

Owner Benefits

Potential owner benefits resulting from the CM organizational structure are as follows:

1. Potential reduction of contractor overhead through the elimination of the general contractor and the "flattening out" of the organizational structure of the construction contractor team.
2. Potential reduction of architect/engineer design fees due to a reduction in their involvement in project estimating and administration.

It can be argued that the above benefits are obtained independent of whether or not the CM firm provides any services. In other words, the CM organizational structure results in owner benefits that partly offset the fee of the CM firm.

The CM structure often results in the elimination of the general contractor. This was discussed in the previous chapter and illustrated in Figure 1.8. Given this elimination, the CM process in effect eliminates the overhead that the general contractor includes in its price.

The savings due to the elimination of the general contractor's markup on subcontractors and the "managing" fee are somewhat difficult to quantify. However, based on the premise that the general contractor subcontracts 50 percent of the construction work, and the fact that the subcontractors mark up their work 20 percent for overhead and profit, and the general contractor in turn duplicates the subcontractors' overhead, the potential savings from the elimination of the general contractor on a proposed $1 million contract can be calculated as follows.

General contractor award	$1,000,000
Subcontracted work	500,000
Subcontractors' markup (20%)	100,000
General contractor's subcontractor markup	100,000

The $100,000 general contractor's subcontractor markup is potentially delivered to the owner through the elimination of the general contractor and the use of separate contracts to independent construction firms, as illustrated in Figure 1.8. Admittedly, the competitiveness of the construction industry may prevent the general contractor from doubling up on the markup of all of its subcontractors. However, even if one assumes as small as a 5 percent general contractor markup on subcontractor bids, a $25,000 savings (one-fourth of the $100,000) results from the elimination of the general contractor markup. This $25,000 equals 2.5 percent of the total budgeted construction contract amount. Given the fact that the CM firm may charge in the range of 2 to 8 percent of the contract amount for its services, the potential savings from elimination of the general contractor are significant.

It should be noted that the CM process organizational structure does not always eliminate the general contractor. An organizational structure that included the CM and general contractor was illustrated in Figure 1.9. When this is the case, it can be argued that the sum of the general contractor's overhead is actually duplicated by the CM firm, rather than reduced.

Also, even when the CM process eliminates the general contractor, the saving in the general contractor's markup is offset by the addition of the CM firm's fee. In other words, the CM firm is being paid a fee for performing some of the functions previously performed by the general contractor. However, the CM process is intended to do more than merely change the name of the firm in charge of the project. The CM firm is expected to perform services that exceed those normally expected of the general contractor. Services performed by the CM firm are set out in chapter 3.

As we will discuss in chapter 3, the CM firm assists the project design firm throughout the design phase of the project and usually relieves the design firm of some or all of its construction supervision and inspection functions. During the design phase, the CM firm serves as an adviser of the availability and cost of construction material, labor, methods, and subcontracts. The CM firm also has the responsibility for preparing project cost estimates throughout various design phases. In the non-CM process, the project designer has the responsibility of preparing project design estimates for the owner.

Given the fact that the design firm is almost totally relieved of estimating, supervision, and inspection duties and is also assisted in its design duties through the injection of CM knowledge of cost and availability of materials, labor, and contractors, it follows that the fee charged by the design firm in a CM process should be reduced from the fee charged in a non-CM process.

Architect/engineer design fees (including the fee for estimating, design, and defined inspection and supervision services) range from as little as 3 percent to as much as 12 percent of the cost of a project. Eight percent is an average design fee. The majority of the design firm's fee is for its preparation of project contract documents, including drawings and specifications. However, one has to assume that part of its fee includes estimating and supervision services. Assuming this fee is as small as 1 percent (one-eighth of its total fee) of the projected project cost, this fee equates to $10,000 on a $1 million project. The use of the CM process may eliminate this cost because the CM firm performs and gets paid for its estimating, inspection, and supervision services.

In practice, many architect/engineer design firms continue to charge the owner the fee charged in a non-CM process, when performing in a CM process. This is partly the result of not being able to single out the design firm's fee for specific services such as project inspection.

On the surface it appears that the design firm would resist the CM process because the potential reduction of the firm's fee. However, it is very important to note that the actual design of the project is left to the architect/engineer design firm. In effect, the CM process places the design firm in the role it is trained to perform and is most comfortable with—the preparation of project drawings, specifications, and contract documents. In effect, the CM process removes the design firm from its riskier and less profitable functions—estimating, project inspection, and supervision.

The potential reduction of contractor overhead and the possible reduction in the fee of the design firm does not in itself result in the CM process being beneficial for a potential project owner. It is the services of the CM firm and the organizational structure in which they are performed that yields the project owner significant benefits. The following are some of these benefits.

Systems approach to project The systems approach to management states that the optimal solution to any problem can be achieved only if all of the relevant factors and components of the problem are related and analyzed. In effect, the systems approach says that one cannot optimize the building of a construction project unless one interrelates all of the parts. The fact that the CM process has a single firm involved throughout design, construction, and implementation, places the CM firm in a position to interelate all relevant variables, and to make decisions that minimize the project's time and cost and maintain or maximize the project quality.

Construction knowledge input into design As noted previously a project design cannot be optimal unless it includes input regarding the cost and availability of labor, materials, subcontractors, and the feasibility and economics of the proposed construction methods. Knowledge of these parameters is obtained from construction experience. Because the CM firm is expected to possess such experience and because the CM firm should be engaged to work with the project's architect/engineer during the design phase, the CM process merges the design skills of the architect/engineer design firm with the construction (and design) knowledge of the CM firm. This merging of skills and knowledge enables the optimal design of the project regarding time, cost, and quality.

Estimating knowledge introduced into design The design for a proposed project evolves over several weeks, months, or even years, depending on the size of the project. During this period, several levels of estimates are prepared for the owner by the designer. The size and quality of the project designed are dependent on the designer's conception of the cost of building the project. In effect, the various cost estimates prepared by the designer help to dictate the project design.

It can be argued that the design firm's lack of everyday involvement in the construction industry makes it unable to make dependable and reliable estimates. There is no doubt that if the estimates are grossly incorrect, both the quality of the project and the time it takes to complete a project suffers. For example, if the designer estimates a project to cost the owner $1 million, but in fact contractor bids totaling $1,200,000 are received, the project may need design modifications that could lessen the project quality and more than likely will result in extended project design time, which in effect results in additional overall project time.

The introduction of the CM firm into the design process results in more accurate preconstruction estimates. The supplementing of the designer's skills with the construction cost information knowledge of the CM firm should enable estimates to be presented that improve the project design and also prevent project time delays resulting from duplicate design or project letting iterations.

Single source of management A design firm is engaged primarily by the potential project owner for its design skills. The contractor is engaged primarily for its skills with "bricks and mortar," in other words, its ability to build. It might be argued that apart from the CM process, the project owner has no entity to manage its project. Although it is true that the contractor implements management practices, including estimating and scheduling, it does this out of self-interest. The CM firm applies its management skills (each of which is described in a following chapter) throughout the project phases, with the objective of serving the project owner. In effect, the CM firm provides the owner an independent, objective manager working in the behalf of the owner from the start to the conclusion of the project.

The introduction of management skills into the proposed project complements the architect/engineer's design skills and the contractor's building skills. Without the management input, both design and construction may be performed suboptimally. For example, just because a contractor is working at the job site with little idle labor time does not mean that the construction is being performed efficiently. The introduction of management introduces "work smarter not harder" into both the design and construction phases of a project.

Reduces project owner's time and involvement The time commitment of a project owner during planning, design, and construction phases of its project in the non-CM process varies, depending on the services of the designer and contractor teams. However, it is rare that the project owner is not called upon to play a significant role in the decision making throughout the process. For example, the project owner will likely spend considerable time requesting or approving change orders, approving contracts to be signed by contractors, requesting and monitoring project financing, and serving the reporting and legal needs of external parties, such as government agencies. The need to be involved in the building process has prompted some project owners to maintain a staff of individuals (usually engineers, architects, or construction supervisors) to enable required decisions and time commitments to be reached. This is especially true for manufacturing companies that may have an almost constant building process. Although this in-house staff may perform required owner functions, often their services are duplicated by the design or contractor teams. Another potential disadvantage of an in-house staff is that the firm's building program may fluctuate and therefore the in-house staff becomes nonproductive and an unwarranted overhead to the owner.

The CM process has the potential to minimize the project owner's time commitment and need for an in-house staff. As an agent of the project owner and given a close and trustworthy relationship, the CM firm should be able to make most of the decisions for the owner once the scope of the project and budget are established. The aggressive CM firm will handle such functions as project planning, the initiating and monitoring of change orders, the selection of project contractors, and duties regarding the financing and monitoring of the owner's cash.

In practice, most project owners want some involvement in the building process. The CM process does not prevent this involvement, and can be structured to enable as much or as little owner involvement as the owner prefers and/or needs.

Phasing of work to reduce project time schedule Although several different types of owner/contractor relationships exist, the most common is the process whereby the owner engages a general contractor for total project responsibility and the general contractor in turn awards work to several subcontractors. Because the owner only signs a contract with the general contractor, the owner prefers the general contractor to sign a contract that guarantees the total project cost. Given the need for the general contractor to give a total dollar contract commitment for the project, it follows that the entire project drawings and specifications must be prepared before the general contractor's lump-sum bid is obtained. This means that it may not be possible to start any construction until the project drawings and specifications are complete. It is true that the owner might have the contractor start construction before the designer's completion of the drawings and specifications if it awards the contractor a time-and-material or cost-plus contract. However, these two forms of contracts are often criticized because they may not promote cost savings or productivity incentives for the contractor.

A project owner's interest in being able to start construction before drawings and specifications are complete centers on being able to condense the overall project schedule. On occasion, the duration of the project design phase may equal or exceed the construction phase. Months of project time might be eliminated if part of the construction is initiated during the design phase. This is illustrated in Figure 2.1. For example, the project excavation, substructure, and possibly structural frame may start or be completed while the "finishers" and specialty systems are being designed. Other advantages, and possible disadvantages, of phasing a project's construction are discussed in chapter 9.

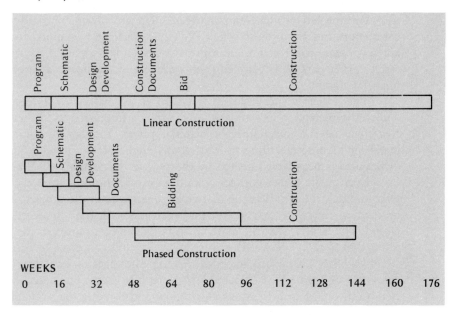

Figure 2:1 Linear construction versus phased (or fast-track) construction

Whereas the organizational structure of the general contractor construction process may make it difficult for the owner to phase the construction of a project, the CM process and the manner in which contracts are awarded facilitate the phasing of the project's construction. As illustrated in Figure 1.8, the construction of a CM project is normally performed by means of separate contracts. Each contractor has a direct contract with the owner and is engaged to perform a well-defined portion of the total construction work. For example, separate contractors may be engaged for the excavation, concrete work, masonry, structural steel, electrical work, and so on. Because each firm contracts directly with the owner, it is feasible for the owner to receive bids and performance from some of the early-phase contractors (for example, excavation, substructure, concrete) before later types of construction are designed. In effect, the use of separate owner/contractor contracts enables the phasing of the design and construction phases of a project.

Enable owner-purchased materials Material shortages and rapidly escalating material prices have become constraints on the project owner's ability to minimize project time and cost. All too often the inability to have required material delivered to the project when needed has resulted in lengthy project delays. These delays have a "rippling" effect when it comes to contractor performance, because the performance schedule of each of the project contractors is dependent on the schedule of each of the other contractors.

Unexpected or uncertain material costs are detrimental to both the contractors and the project owner. When a contractor commits to a dollar amount before construction, unexpected material costs may be incurred. The result is that part of its planned profits are lost or the firm may even lose its financial stability.

The project owner indirectly pays for rising material costs because the contractor includes in its estimate and contract a forecasted and a contingency cost to offset the uncertainty of material prices. Should the contractor go bankrupt because of failure to forecast material costs accurately, the project owner also is negatively affected by the project delays that will develop.

Even if the owner engages a contractor on a time-and-material or cost-plus contract, the fact that the material is purchased as the project progresses means that the owner pays a cost that increases with time in an inflationary period. The result is that the material costs for the project exceed the material prices that were in effect at the time the project was conceived or designed.

To offset the consequences of material shortages or escalating material costs, the CM process is characterized in part by owner-purchased materials. Because the CM firm is expected to know and forecast the availability and cost of materials, it may have the owner purchase some materials before the contract to install them is awarded. For example, if the project design is to the point where the structural steel frame is fully determined, and the CM firm recognizes that structural steel is forecasted to be in short supply or expected to rise in cost significantly, it may prevent a project delay as well as be able to have the material purchased at a lower price by having the owner purchase the material rather than the subsequently engaged structural steel contractor.

Assume that there is a six-month delivery schedule for structural steel and it is anticipated that a structural steel contract will be awarded six months after the time the structural steel is designed. Let us also assume that the price of the required structural steel at the time of design is $100,000 and is expected to increase at 1½ percent per month for the immediate future. Figure 2.2 illustrates the project owner's potential savings in time and cost if the structural steel is purchased directly from the steel fabricator at the time it is specified in the design process.

The fact that the project owner purchases the material at the time of design does not necessarily mean that the owner has the material delivered to the project site then or even pays for the material at that time. It merely means that the owner signs a commitment letter with the material vendor to have delivered an agreed amount of material at a defined point in time and at an agreed-upon price. Although it is true that the dollar savings illustrated in Figure 2.2 may be offset somewhat by the fact that the material vendor may hedge its price to reflect future costs, there may also be owner savings from the elimination of the project delay if the contractor were to order the

Benefits of the CM Process 31

	Savings	
1. Purchase Cost:		
Contractor supplied	$100,000	
Inflation		
0.015 × 6 × 100,000	9,000	
	$109,000	
Owner supplied	100,000	$ 9,000
2. Elimination of Contractor Markup:		
Assume 2.5%—0.025 × 100,000		2,500
3. Elimination of delay		
Assume 1½% per month delay		
9% × 100,000		9,000
Potential Savings of Owner-Purchased Material		$20,500

Figure 2:2 Benefit of owner-purchased material

material, and savings that may result because the owner, due to a better financial position, may be able to negotiate a better price for the material.

It is true that a designer in a non-CM process may assist the owner in identifying material that can be advantageously purchased by the owner. The CM process is not unique in this regard. However, the CM firm's knowledge of the construction industry enables a better execution of owner-purchased material than in the non-CM process.

Elimination of adverse relationships Perhaps the most important benefit of the CM process is the potential reduction or elimination of the adverse relationships that characterize the non-CM building process. All too often in the non-CM process, there are several adverse relationships among the firms involved in a project. For example, the relationship between the project designer and contractor team is often strained. The contractor sometimes takes advantage of the designer, and vice versa. For example, the contractor, when performing its estimating function, may not inform the designer of an obvious design omission because the omission may subsequently result in a beneficial change order for the contractor. Equally important, the designer may take advantage of the contractor by forcing it to perform some work at the job site for which it is not compensated. The work may stem from a design omission. The contractor cooperates with this "do work for no compensation" process because the designer may hold back payment if the work is not performed.

The CM process emphasizes team work. It has the objective of having each of the entities involved work with rather than against each other. The manner in which the CM firm achieves this teamwork is not easily defined or quantified. However, its ability to do this has a major impact on its ability to deliver the project owner, project time, cost, and quality. It can be argued that if the CM process, through the CM firm does not develop a teamwork process and defeat the potential adverse relationships that can exist among the project entities, that the project's time, cost, and quality may be worse than the non-CM process.

The means by which the CM process accomplishes a teamwork environment is based on the communication skills of the CM firm and its ability to lead and to gain the respect and cooperation of the entire building team. However, the CM organizational structure may itself produce more teamwork than the non-CM process.

External Party Benefits

Although a potential project owner uses the CM process for anticipated benefits, the potential benefits or disadvantages of the CM process for external parties should also be of concern to the owner. If the CM process is not compatible with the interests of external parties, including the project's design and contractor teams, their unwillingness to cooperate will lead to the demise of the CM process or, at a minimum, fewer benefits for the project owner.

Fortunately, the CM process can lead to benefits for both the design team and the project's contractor team. The architect/engineer design team is typically relieved of the project supervision and inspection duties when the CM process is utilized. The project owner usually contracts with the CM firm to perform these services. Traditionally many architectural or engineering design firms have not felt comfortable with its project supervision and/or inspection functions. The firm's dislike of these functions is related to the potential liabilities associated with them. In general, the architectural or engineering design firm has attempted to be "held harmless" in its business dealings.

The design firm's dislike of project supervision and inspection also relates to its lack of training and experience to perform this function. The design firm is trained to design, not manage. The use of a CM firm places the design firm in a position of only performing its primary skill, that of a designer.

The CM process also relieves the design firm of some of its project budgeting and estimating duties. The CM firm assists the design firm in these functions by offering its knowledge of costs and availability of materials, labor, and contractors. The architectural or engineering design firm often feels uncomfortable with the budgeting or estimating functions because its lack of everyday involvement with the construction project does not provide it with the necessary knowledge to prepare accurate project budgets or estimates. In

Benefits of the CM Process

effect, the involvement or assistance of the CM firm with these functions should result in the design firm being more comfortable with project budgeting and estimating.

The construction firm's acceptance of the CM process varies, depending on the type of construction the firm performs and its attitude regarding its role in the construction process. There are several potential benefits to the contractor resulting from the CM process. If the CM firm performs its management duties, the contractors should be better able to perform their construction work. The CM's active involvement in project scheduling and control, along with its project organizing procedures and communication, including job-site planning meetings, should provide the contractors a productive work environment that enables them to attain their planned profits and minimize construction delays.

A second potential benefit to contractors involved in the CM process is that their direct prime contract relationship with the project owner should enhance their receipt of progress payments. A subcontractor to a general contractor in a non-CM process may receive its owner-held retainer only after the general contractor receives its retainer. Given the fact that the general contractor may not receive its retainer until the project is substantially complete, some subcontractors may not receive their retainers until well after they have completed work and perhaps are no longer even at the job site. The result is a potential cash-flow problem for the subcontractor, as illustrated in Figure 2.3.

The organizational structure of the CM process enables the releasing of individual contractor owner-held retainers as they complete their phase of the construction process. Because there is no middle man between the contractor and the owner, the individual contractors receive any owner-held retainers earlier than in the non-CM process.

Not all contractors would agree, but it can be argued that the contractor's prime contract relationship with the project owner in the CM process, in addition to speeding up its receipt of progress billings and retainer, also provides the contractor a better communication channel to the owner, with the result that the contractor is in a position to obtain quicker decisions that facilitate production. In other words, the elimination of the middle man removes the possibility that the specialty contractor is left in the dark as to what to do next.

The above argument is not accepted by all contractors. Some specialty contractors feel that the inclusion of the general contractor provides needed leadership. They argue that without the general contractor's involvement, the specialty contractors may be left in a position of not knowing what to do next. In other words, some owners, along with their CM agents, may not provide the needed leadership, with the result being confusion and project delays, overlapping jurisdictions, and possible omissions. Obviously, the question as to whether or not the contractor gets needed leadership from the project

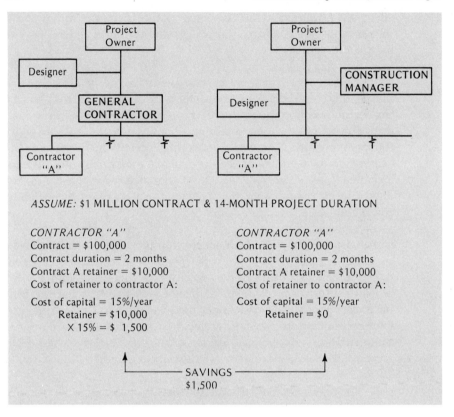

Figure 2:3 Impact of separate contracts on retainer

owner or CM firm is dependent on the skills and services of the CM firm involved. If the CM firm performs the services it contracts to perform, that of being the project leader, the individual contractors should benefit from the CM process and the prime contract relationship with an owner.

DISADVANTAGES OF THE CM PROCESS

The CM process is not without its disadvantages. The potential project owner must evaluate the potential advantages and disadvantages when deciding on the feasibility of using the CM process. Because both the advantages and disadvantages are in part dependent on the scope and size of the proposed project, the project owner must critique each project by itself when evaluating the feasibility of the CM process. Potential owner and third-party advantages accruing from the CM process were discussed in the previous sections of this chapter. Potential disadvantages are discussed in the following sections.

Owner Fee to CM Firm

An obvious disadvantage of the CM process is the additional fee the project owner pays, over and above that paid to the project designer and contractor team, to the CM firm for its services. Although it is true that the CM process may offer the project owner dollar benefits exceeding the fee paid to the CM, the fee in itself has to be viewed as a potential disadvantage. This is especially true in light of the fact that the dollar benefits from the process to an owner are unsure and may not be able to be quantified at the time the owner commits to the CM process.

The fee paid to the CM firm may be in the form of a fixed fee, a combination of a fixed fee and reimbursable items, or as a percentage of project cost. If the project's scope and size are well defined, the preferable fee arrangement between the CM firm and the project owner is a fixed fee. The fixed-fee contract results in the fewest distractions or incentives for the CM firm to serve anything but a true agency role. In a fixed-fee agreement, any significant change in project scope or size results in the need for the agreed-upon fee to be adjusted upwards. In addition to the fact that the fixed-fee arrangement may "best fit" the agency relationship, it can be argued that the CM firm in holding itself out to have estimating skill (one of the desirable traits of a CM firm) should be able to estimate the cost of its own services for the project owner.

In practice, many CM firm/project owner agreements result in the CM firm being paid a fixed fee plus reimbursables for items defined in their agreement. The most common reimbursable in a CM contract is supervision. This is based on the fact that it may be difficult for the CM firm to establish the required number of on-site supervision hours before or during the project. Instead of including a fee for on-site supervision in its fixed-fee amount, the CM firm may quote an hourly fee and bill the project owner for the recorded on-site hours it supervises the project. Although handling supervision on an hourly basis might prove to be workable, it can be argued that because the CM firm has a profit built into an hourly supervision or reimbursable rate, the firm has an incentive to spend unnecessary or excess supervision hours on a project. In effect, the CM firm may act in its self-interest, which is not consistent with its agency agreement.

A CM firm/project owner agreement whereby the fee is based on a percentage of project costs shares all of the disadvantages of a cost-plus contract for engaging contractors. A cost-plus agreement may promote inefficiencies and lack of motivation. In regard to the CM firm, a cost-plus agreement results in added compensation as the project cost increases, while in fact the CM firm is engaged to minimize the project cost.

The total fee paid by the project owner to the CM firm should vary as a function of the services provided by the CM firm as well as the complexity and size of the project. Because of at least these three variables, services, project complexity, and project size, it is difficult to set out a typical fee. However, in practice the compensation paid to the CM firm ranges between 2 and 8 percent of the total cost of the project. Figure 2.4 illustrates typical CM compensation as a function of total construction costs for different types of CM services provided, project complexity, and size. It should be noted that the fees should at best be viewed as a guide. The fee paid to the CM firm should be established through a detailed analysis of the project and the services to be performed.

		Project Size		
		$1 million	$1–5 million	$5 million or more
Type of CM Services	Pre-Construction	3–4%[a,b]	3–4%	2–3%
	Construction	4–5%	3–5%	3–4%
	Pre-Construction and Construction	6–8%	5–7%	4–6%

[a] Percentages are CM fee as a percentage of total project cost.

[b] First % shown represents low project complexity; second % represents high project complexity.

Figure 2:4   Suggested CM fees

Requirement for Bonding of Individual Contractors

Advocates of the traditional non-CM process point out that one of the potential disadvantages of the CM process is less contractor competition for contracts relating to the need for individual contractors to be bonded. In the non-CM process illustrated in Figure 1.7, it is common for the general contractor to provide any project-owner-requested bonds for the entire team of construction contractors. For example, the general contractor often provides the performance bond for its subcontractors, including electrical, mechanical, and plumbing specialty contractors. In effect, the specialty contractors are freed of the need to obtain bonding because they come under the umbrella of the general contractor.

It can be argued that the owner benefits from the process in which the general contractor provides bonds for the subcontractors (that is, the non-CM process). The general contractor providing the bond enables the nonbondable specialty contractors to perform work at competitive prices. The owner receives protection against subcontractor failure through the general contractor's

Disadvantages of the CM Process

bond, and also obtains the benefit of specialty contractor competition, which leads to lower specialty contractor bids.

The most common organizational structure for the CM process is illustrated in Figure 1.8. Because each contractor has a prime contract with the project owner and also because the CM normally does not bond the planned construction work, any bonding requirements must be the responsibility of the individual contractors. Each contractor must obtain its own bond.

Given the fact that many specialty contractors have relatively weak balance sheets, with relatively little working capital and insufficient equity, it may be difficult for them to obtain bonds. Thus specialty contractors that have the know-how to perform the necessary construction work may be unable to bid due to their inability to obtain bonding. The result may be less specialty contractor competition with a less competitive price for the bid packages. The opponents of the CM process argue that the need for individual prime contractors to provide their own bonds will result in only the larger specialty contractors being able to bid.

Although it is true that the use of separate prime contracts may decrease specialty contractor competition, it is likely that this disadvantage is overemphasized. For one, the disadvantage relative to the non-CM process diminishes when one recognizes that the potential project owner may request bonds from individual subcontractors even when utilizing the general contracting process. There is a trend for the owner to request individual contractor bonds to protect against subcontractor financial failures.

The potential negative impact of individual contractor bonding requirements is also lessened by the fact that the engaged CM firm is in a position to evaluate contractor risk and may in fact wave the bonding requirement for a given contractor. This is an acceptable practice in the private sector, (public construction laws, such as the Miller Act, may require bonds to be secured for all phases of the construction work). For example, the CM firm may, through past experience, know that an electrical contractor is reliable, in spite of its inability to obtain a bond. The CM firm may recommend that the project owner waive its bonding requirements and engage the contractor. The result is that the potential lessening of contractor competition is decreased or eliminated.

Responsibility without Authority

Perhaps the most serious disadvantage of the CM process relates to the relationship that the CM firm has with the selected construction contractors. The CM firm in Figure 1.8 is in the classic worst-management position, that of responsibility without authority. The CM that contracts with the project owner to deliver project time and cost, but without direct contract with the construction contractors may not have authority to force the contractors to perform to the owner's advantage.

It is true that the CM firm, acting as the project owner's agent, indirectly has all of the authority of the owner. However, the fact that the CM firm is not involved in a construction contract with the selected contractors (as would exist with the general contractor/subcontractor process) may in fact result in an adverse or nonworkable CM firm/contractor relationship.

The fact that the CM firm normally approves payments to engaged contractors gives it some authority over the contractors. However, this type of authority can be viewed as "negative management" and is not usually effective long-term management. In other words, authority gained through holding back payment will likely be met with contractor discontent and possibly the contractor's striking back by actions or nonactions such as knowingly understaffing the construction project, leading to project delays.

The potential disadvantage of the CM firm not having a productive working relationship with the project's contractor should be of concern to the project owner. The CM firm markets its ability to secure contractor interest in bid packages and, once contractors are contracted, to develop a team concept whereby all entities work toward the common goal of delivering the project owner, optimal project time, cost, and quality. These are the expected services of the CM firm, but the lack of a productive CM firm/contractor relationship can result in the reverse.

Contractors may view the CM firm negatively. This is true if the CM firm takes the position of "we know it and you don't." A CM firm attitude of talking down to potential contractors rather than working with contractors will likely lead to inadequate contractor bidding and noncooperation and disruption throughout the construction phase of the project. This potentially dangerous relationship is made worse by the CM firm's lack of contract (and the authority implicit with a contract relationship) with the project contractors.

The means of successfully defeating a nonproductive CM firm/contractor relationship and the CM firm's lack of authority is the CM firm's development of a "work with" relationship with the contractor. The CM firm must promote the concept that it is trying to help the contractors by providing management services. In effect, the CM firm is marketing the fact that it can deliver its services to the owner only if it helps the contractor team and vice versa. While it holds itself out to be skilled in managing time and cost, the CM firm must also recognize the building expertise of the contractor team.

If the CM firm is successful in soliciting the help of the contractor team, the firm will gain authority over the contractors by the choice of the contractors. In other words, the contractor team that respects the CM firm will follow the directives of the CM firm, even if it does not have a contract. For example, assume a CM firm aids one of the project contractors through a recommendation regarding better use of the contractor's equipment. Subsequently, the CM firm requests the contractor to double its crew size to enable the contractor's work schedule to be shortened thus condensing the project schedule. More than likely, the contractor, remembering the previous favor will oblige. In effect, the CM firm has obtained authority without a contract relationship.

It can be argued that a contract relationship never really gives the type of authority needed to successfully execute a construction project. Authority is obtained by positive actions and by a mutual respect of the parties involved. If one accepts this, then the CM firm's lack of contract relationship with the project's contractors should not be of concern to the project owner or the critic of the CM process.

WHEN CM IS BENEFICIAL

A potential project owner has several alternatives when determining how to execute its planned project. The most obvious are the general contractor process and the CM process. The benefits of the CM process relative to the non-CM process vary depending on the project and the needs and skills of the owner. It is therefore necessary that each project be evaluated individually when determining the feasibility of the CM process.

There are three independent considerations that should be analyzed by the potential project owner when evaluating the feasibility of the CM process.

1. Project size measured in dollars
2. Project complexity
3. Project owner's in-house skills

Usually the larger the project size, the easier it is for the CM firm to provide benefits that exceed the firm's fee. This is not to say that a relatively small, complex project does not justify the CM process. However, the mere fact that the "cost of learning" a project decreases as the project cost increases better justifies a fourth party.

It is probably true that project size receives too much emphasis in regard to the feasibility of the CM process. In the early years of CM, many individuals argued that a project had to be in excess of $5 million to justify the CM firm's fee. However, as the CM concept has grown in popularity and sophistication, the dollar value of the projects on which the CM process has been utilized has decreased. Owners of $1 million projects or less have utilized CM.

Project complexity should be a major consideration when evaluating the feasibility of CM. As the project increases in complexity, the benefits of the CM process increase. Projects that involve unique designs or complicated construction result in the CM firm being able to yield the owner significant benefits.

The relationship between the benefits of the CM process and the complexity of the project is undoubtedly related to the fact that the CM process is most used for building and industrial projects. These projects tend to be more complex than residential or heavy and highway construction projects.

Each building or industrial project tends to be unique. Such projects as office buildings, hospitals, schools, and nuclear plants tend to be one of a kind. In addition, each of these projects usually entails the installation of numerous types of materials and systems. Several labor crafts and specialty contractors are employed at the job site. Given the numerous project components and the need to interrelate them into a completed project, the need for the management skills of the CM firm increases. On the other hand, projects that are well defined, with very streamlined and repetitive design and construction phases, lessen the potential benefits of the CM process for the project owner.

The feasibility of the CM process is also dependent on the in-house skills of the potential project owner. As noted in an earlier section of this chapter, one of the advantages of the CM process is that it can reduce or eliminate a project owner's need for an in-house staff to coordinate the proposed project. In effect, the CM firm, acting as the project owner's agent, can perform all the decision-making duties necessary to the successful completion of the project.

If a potential project owner has an in-house staff of designers and construction-skilled employees, the skills and services of the CM firm may overlap or prove redundant. For example, some large manufacturing firms, due to the fact that they have an almost ongoing building program, may employ several in-house employees capable of performing some of the services of the CM. Given these conditions, it may be difficult for the CM firm to justify its fee. In effect, the fewer in-house skills the potential project owner has, or the less it is able to utilize these skills for its proposed project, the more advantageous is the use of a CM firm and process.

The project owner must consider all three factors, project size, complexity, and availability of owner in-house skills to evaluate the impact of the CM process. The combination of a relatively large project of high complexity, coupled with an owner with little or no in-house expertise, is likely the best CM project. However, even the introduction of one of the considerations, for example a complex project, may justify the use of the CM firm in certain cases.

EXERCISE 2.1

BENEFITS OF CM TO SPECIFIC PROJECTS

Several benefits associated with the use of the CM process were outlined in this chapter. These included a systems approach to the project, the benefit of introducing estimating knowledge into design, and establishing a single source of management. These and the other potential benefits vary depending on the type of construction project being considered. Assuming these types of projects: a) residential, b) commercial and industrial, and c) heavy and highway, evaluate each benefit outlined in the chapter relative to its potential impact on these types of projects. Rate the benefit a 3 for having the most favorable impact, a 2 for some favorable impact, and a 1 for having little or no impact. For example, you might rate the benefit of a systems approach a 3 for commercial and industrial projects, a 2 for residential projects, and a 1 for heavy and highway projects.

EXERCISE 2.2

SELECTING A CM

This chapter discussed the fact that a potential project owner has an option between engaging a general contractor or a construction manager. Also, if it judges a construction manager appropriate, based on information in chapters 1 and 2, the project owner still has an option of requesting a guaranteed maximum CM contract or a nonguaranteed maximum CM contract. Based on the considerations discussed in this chapter, prepare a suggested set of procedures in the form of a decision flow chart to aid a potential project owner in choosing among a general contractor, a construction manager with a guaranteed maximum contract, or a construction manager with a nonguaranteed maximum contract.

3

CM Services and Forms of Contracts

ENGAGING A CM FIRM

The contract agreement between the project owner and the CM firm dictates the services and the liabilities of the CM firm. Since every project owner requires a different set of services, and almost every CM firm possesses a different set of skills, the project owner/CM firm contract agreement should be a custom designed document for almost every project.

Although each project justifies a custom-designed set of CM services, in practice the project owner and CM firm often sign one of two standardized forms. Both the American Institute of Architects (AIA) and the Associated General Contractors of America (AGC) publish standard CM contracts. These standard forms are illustrated in appendixes A and B. Differences between the two documents are discussed in this chapter.

The use of standardized project owner/CM firm contract agreements has led to project owner acceptance of the contract without a detailed analysis of the services to expect or its liabilities and the liabilities of the CM firm. Equally troublesome is the fact that the CM firm offers the project owner one of the standardized contracts without the ability or the intent to deliver all of the services set out in the document. Any standardized form can serve a negative purpose if both of the parties to the agreement are unaware of or misinterpret the contract provisions.

Although the use of a standardized agreement can prove detrimental if it is taken lightly or misinterpreted, it can also prove as a good base for custom designing an agreement. The vague nature of CM and the services to be performed by the CM firm make the preparation of a CM agreement tedious, time consuming, and critically important.

Each and every project owner/CM firm agreement should contain certain basic sections, including core services and owner/CM firm relationships. Services over and above the core services should be identified by the project owner and CM firm and set out in the contract. Both core CM firm services along with other CM services are discussed in this chapter.

KEY SERVICES OF THE CM FIRM

The services of a CM firm vary in practice from a few services yielding the project owner few benefits to a broad set of services ranging from assisting the owner in project feasibility calculations to property management of the completed project. Although the services vary, there tend to be a few "buzz word" services that are identified with the CM process. The following might be considered as the most common services of the CM firm.

- Project feasibility and tax analysis
- Parameter estimating
- Value engineering
- Project planning and scheduling
- Phased construction
- Packaging construction work
- Project control information system

A more detailed list of specific CM services is outlined in the following sections of this chapter; however, the objective of each of the key CM firm services is stated in the following paragraphs. Chapters 5 through 10 discuss these key services in more detail.

The objectives of the CM firm assisting a project owner in evaluating the project feasibility and tax analysis are twofold. For one, the CM firm adds valuable information to the feasibility calculations. Without the correct input of construction costs and ongoing expected maintenance and repair costs, the validity of any project feasibility study is questionable. To this extent, the CM firm's involvement adds credibility and validity to the process.

Utilizing its accounting, tax, and construction skills (assuming the CM firm has these skills), the CM firm is in a position to assist the project owner and the design team in maximizing potential tax savings by relating the project design to componentized depreciation and investment tax credit analysis. Both the project's feasibility and design should recognize the impact of these factors. Project feasibility calculations along with componentized depreciation and investment tax credit analysis are discussed in chapter 5.

The importance of parameter estimating to the CM process relates to the critical nature of being able to establish accurate cost estimates of the anticipated project while the project is in the design phase. The scope of the project, including both the project size and quality are dependent on the designer's conception of project component costs. If this conception is in error, then the project design will be less than optimal. For example, if the design is based on an anticipated cost that proves to be less than subsequent contractor bids, the project designer will likely have to redesign the project, which will lead to delays and to a scaled down project that is inferior to what would have been programmed into the budget.

Parameter estimating, along with other CM estimating techniques, has the objective of enabling accurate design estimates. The technique, along with the input of construction knowledge by the CM firm, provides the potential for improved design estimates. Parameter estimating and other CM estimating techniques are discussed in chapter 6.

Value engineering is not a management technique unique to the construction industry. It is a tool for optimizing the design of a physical entity or process. In regard to the CM process, value engineering provides the CM firm a means of inserting its knowledge into the designer's proposed contract documents.

Looked at another way, value engineering enables the CM firm to evaluate the project design with the objective of optimizing the time, cost, and quality of the completed project. Value engineering is presented with applications in chapter 7.

Project planning and scheduling are perhaps the most basic of all CM firm services and are critical throughout all phases of a project. Numerous project planning and scheduling techniques are available to the CM firm.

Project planning and scheduling serve two broad purposes in a construction project. For one, the preparation and analysis of a project plan and schedule provide the CM firm a means of modeling the project to minimize project time and cost. There are numerous means of obtaining each and every phase of a project. The preparation of a plan and schedule enables the determination of the optimal means of accomplishing the owner's time and cost objectives.

A project plan and schedule also serve a project control function. Control of time and cost is in part attained by comparing project progress with the prepared plan and schedule. Without the plan and schedule against which progress can be measured, the ability to control design time, project labor and material expenses, and contractor performance is decreased if not eliminated.

Project planning and scheduling, in addition to the two major purposes of planning and control, serve yet another support objective. The CM firm's ability to perform two other key CM services, phasing and packaging construction work, is made possible in part from the preparation of a project plan and schedule. As will be illustrated in subsequent chapters, the very question as to what to phase and what to package and the benefits of both of these services are in part dependent on the project plan and schedule. Project planning and scheduling are discussed in chapter 8.

Phased construction is also referred to as *fast-track* construction. The objective of phasing the project is to merge some of the project design, letting, and construction phases to shorten the project time. A decreased project time will also result in a decreased project cost. The potential benefits of overlapping project phases, for example, starting substructure construction work while drawings for the project finishes are in process, can be partially offset by design omissions and subsequent change orders if the phasing of a project is not performed adequately. The means and justification, and potential benefits of phased construction are discussed in chapter 9.

Packaging of construction work relates to the CM firm's segmenting of the total project into distinct packages to be awarded to the prime contractors. This is in contrast to the non-CM process, in which a single contract is typically awarded to a single contractor, the general contractor.

The number of contract packages in a project varies. It may range from as few as five or six to as many as forty for a complex, large project. The objectives of packaging work are twofold. One, in packaging work, the CM firm attempts to segment individual packages to yield competitive bids from contractors. In theory, the CM firm seeks the combination of packages that yields the minimum project cost. A second objective of packaging is to achieve a project schedule that minimizes project construction time. For example, the construction manager may be able to decrease or minimize the time associated with placing concrete by awarding three separate concrete contracts to individual firms and having the firms perform their work simultaneously.

To a lesser extent, the packaging of work can attain the objective of satisfying various government or owner contract-awarding requirements. For example, the size of the contract packages that the CM firm sets out will in part determine which contractors can or will be willing to bid on the work. Therefore, it is possible to satisfy minority set-aside programs or promote or discourage out-of-town contractors by means of packaging contracts. Packaging of construction work is discussed in chapter 9.

A project control information system is essentially an accounting system for controlling the time, cost, and quality of performance. Such a system is relevant during all phases of a construction project. A project plan and schedule, by itself, will not guarantee performance. Although the plan and schedule may set the performance goals, it is the control function that ensures performance. A control system implies the use of forms and the collection, processing, and analysis of data. Documentation of events and the incurrence of costs are a necessary ingredient in controlling a construction project's time and cost. The concept of a project control information system, along with example data and forms, is presented in chapter 10.

PRECONSTRUCTION SERVICES

The services of a CM firm are often classified as preconstruction or design services and construction services. Both the American Institute of Architects' and the Associated General Contractors of America's standard owner/CM firm contract sets out both preconstruction and construction services.

Preconstruction services can yield the project owner benefits that equal or exceed construction services. The benefits from construction services tend to get the most owner and layman attention, due in part to their conception of the inefficiencies that typify the construction phase of many projects. It is common to hear criticism of the construction firm's inability to keep the project flowing smoothly, referring to delays caused by poor equipment maintenance, inadequate material purchasing, and the firm's inability to properly schedule labor requirements.

Although it is true that several construction phase inefficiencies can be cited, the project owner's ability to attain project time, cost, and quality objectives is also related to several preconstruction decisions and practices. For example, the time, cost, and quality of the owner's project is partly dependent on the accuracy of the various preconstruction estimates, the scope and details of the prepared contract documents, and even the compatibility of project design with favorable tax laws and regulations. The relationship of the time, cost, and quality of the owner's project to these considerations was discussed in previous chapters and will be discussed again in subsequent chapters.

The importance of preconstruction CM firm services is so great that the absence of these services may result in the CM process being unwarranted for some projects. This potential nonoptimal use of the CM process, that is, the limiting of services to construction services, is of special concern because a project owner may engage a CM firm after the project designer has been engaged. In some instances, this will cause the CM firm to be limited to performing construction services because the designer may already have completed the project's contract documents. When this is the case, the potential benefits from CM are curtailed. Ideally, the CM firm is the first contracted firm on a project. This enables the CM firm to deliver preconstruction and construction services.

Figure 3.1 gives a list of CM firm preconstruction services. The list is broader than the list of CM services set out in either the AIA's or the AGC's standard CM contract agreement forms. These standardized project owner/CM firm forms are presented in appendixes A and B respectively. The extended, broad set of preconstruction services listed in Figure 3.1 emphasizes the total system concept of CM. Each of the preconstruction CM services listed in Figure 3.1 is described in following paragraphs.

Owner Needs Identification Study

Perhaps the most important service performed by the CM firm is setting out the needs of the owner from the project's inception to its completion. The need for this service reflects the fact that every project owner is unique in regard to its proposed project and in the services needed to execute its project.

The CM process is a custom-designed process. Identifying all the services to be performed enables a project to be completed successfully and enables the identification of responsibilities of the owner, CM firm, architect/engineer designer, and contractor team. For example, in identifying that a potential project owner needs professional assistance in securing financing, the project

Preconstruction Services

> - Owner-Needs Identification Study
> - Project Feasibility Study
> - Tax Analysis of Project
> - Marketing Research for Proposed Project
> - Assistance in Obtaining Financing
> - Assistance in Obtaining Permits & Zoning
> - Budgeting
> - Value Engineering
> - Parameter Estimating
> - Scheduling of Design & Pre-Construction
> - Identification of Long-Lead Items
> - Bid Packaging
> - Awarding Contracts
> - Setting Out Operating Procedures & Responsibilities
> - Process Paper Work

Figure 3:1 CM preconstruction services

owner and the CM firm may evaluate the feasibility of the CM firm providing the services versus engaging a consulting firm specializing in finance. Identification of owner needs is fundamental to analyzing who is best suited to provide the services.

Project Feasibility Study

During the early phases of a project, it is of the utmost importance that feasibility studies be made. It is difficult to justify the construction of any project, private or public, unless a quantitative analysis is made of its costs and benefits.

All too often an owner has built a project only to discover that the project was ill-planned in its economics or benefits from the day it was conceived. Projects that cannot be completed because of financial hardships during construction are examples of what can result if accurate feasibility studies or estimates are not made. Once the construction phase is started, the project owner is committed regardless of subsequent success or failure. The CM firm's involvement in the feasibility study for a project and the tools or procedures for performing a feasibility study are presented in chapter 5.

Tax Analysis of Project

One of the lesser-known CM firm services, but one that holds significant potential benefits to the project owner, is the performance of the tax analysis of the project. The CM firm, utilizing its construction estimating, accounting, and tax knowledge can help shape the design of the project with the objective of minimizing the project's time and cost and maintaining if not maximizing its quality.

The design of partition walls versus fixed interior walls or the design of a building element to be a production element versus a nonproduction element have tax implications that ultimately affect the cost of a project. The lack of a single source of construction, estimating, and accounting knowledge has usually resulted in the project owner being unaware of the tax implications of its proposed building design. The CM firm can overcome this inefficiency. The tax implications of a project, along with the means of implementing a componentized depreciation and investment tax credit analysis of a project are discussed in chapter 5.

Marketing Research for Proposed Project

An important component of the feasibility study is the inclusion of project benefits as well as costs. For example, the quantification of the potential rental income for a proposed office building or the potential dollar income for a manufacturer proposing a plant expansion can dictate the project design as well as the owner's decision to construct or not to construct the project.

Project owners have on occasion engaged marketing research consulting firms to assist them in quantifying the monetary benefits of a proposed project. The marketing research firm has a defined process of quantifying the dollar benefits. Included in the procedures it uses is *correlation analysis*.

A CM firm offering a broad set of services may include market research analysis to quantify the dollar benefits of a proposed project. The skills necessary to perform a market research study may not be possessed by every CM firm. These skills are somewhat real-estate based. If the CM firm does not possess the necessary skills, it should have the service performed by another firm and perhaps assist the project owner in engaging the firm. On the other hand, the CM firm that has the skills to perform a market research study will be able to better serve the owner in total, due to its involvement and knowledge of all of the project variables from the first day of the project to its completion.

Assistance in Obtaining Financing

Like other preconstruction services, some project owners will take steps themselves or use non-CM firms to assist them in securing financing. However, no two project owners are alike in regard to their skills in obtaining financing or their ability to obtain adequate assistance.

A CM firm possessing accounting and financial skills and experience may assist the project owner in obtaining optimal financing. A project owner has several sources of money available for both the construction and permanent financing of a project. Even from a single source there are several different means of financing. Given the high cost of capital and the fact that the finance cost can weigh heavily on the total project cost, the CM firm's objectives of delivering project time, cost, and quality can in part be served by assisting the owner in identifying and securing the best financing available.

An owner's need for assistance in securing financing for a project is highlighted by the fact that the cost of financing a project can vary significantly for different sources of money and different types of loans. The fact that many project owners are one-time builders and therefore inexperienced in securing project financing also emphasizes the need to secure professional assistance.

Assistance in Obtaining Permits and Zoning

Numerous construction projects get delayed because of the time required to obtain necessary building permits or rezoning of land. Most of these permits or rezoning are the result of governmental (federal, state, or municipal) regulations. Lack of familiarity with these regulations can often result in unexpected delays that ultimately lead to additional project costs. Laws or regulations such as environmental impact regulations or zoning requirements limiting the use of land can frustrate a project owner, especially one that is unaware of these impediments.

The ability to satisfy permits or zoning requirements with no unexpected or excessive project delays is dependent on both an entity's experience in dealing with the laws and regulations and the ability to recognize the need for satisfying the laws and regulations and scheduling the steps necessary to comply with them.

Given the fact that the CM firm is engaged at the conception of a project, and given its broad knowledge of the building process, including permit and zoning regulations, a CM firm can reduce the time and cost of a project by scheduling and taking the necessary steps to satisfy all relevant project permit and zoning requirements.

Budgeting

Potential project owners normally conceive of a project under one of two conditions. They may either have a defined dollar budget and want to build a project that serves their needs and is within their dollar budget amount, or have a defined project and are interested in building the project for the least cost. More often than not, the former is true.

Given either of the above situations, designing a project to budget or designing a given project to minimize cost requires the need to identify costs with proposed scope drawings throughout the early design (or even before) phases of a project. The accuracy of the budget greatly affects the scope of the project and vice versa.

Fundamental to an optimal design is the determination of what is to be spent for the various aspects of a proposed project. Almost every construction project serves to satisfy several owner needs. For example, although an owner may be committed to building a hospital, the design of the hospital will be dependent on the amount or ratio of cost to be expended for emergency care, operating room service, day-care patients, and other special hospital services. These considerations are part of the budgeting process.

Budgeting is the process of planning cash expenditures by project phase or component. Included in the budgeting process are the setting out of a schedule of project owner expenditures as a function of time and the identification of the source of funds. In this sense, the budgeting function is related to the feasibility study and obtaining project financing previously discussed. Fundamental to the setting out of a meaningful project budget is the knowledge of construction costs. Unlike a contractor's detailed project estimate, the budgeting function must be performed without detailed project drawings. As such, the entity performing the budgeting function must be in a position to correlate construction costs to scope drawings or to visualize a project. The ability to perform this function is in great part dependent on up-to-date knowledge of construction prices and documented data from previous similar projects.

Value Engineering

As noted in a previous section of this chapter, value engineering is one of the key or "buzz word" services of the CM firm. Value engineering is a formalized process by which the CM firm can contribute its knowledge of relevant construction into the design process. The critic of the non-CM process points out that without this introduction of construction knowledge, the resulting project design may be nonoptimal or of poor value to the project owner.

Value engineering was briefly described in the previous section. The process of performing value engineering and examples of value engineering are given in chapter 7.

Parameter Estimating

Like value engineering, parameter estimating is one of the CM firm's key services. The process of parameter estimating, along with examples, is discussed in chapter 6.

Unlike the budgeting service previously noted, parameter estimating is usually much more detailed as to expected costs per construction contract to be awarded. Budgeting can be viewed as an earlier-phase estimate than parameter estimating. In a sense, budgeting can be thought of as setting out a project estimate based on functions the project is to serve. Parameter estimating, although somewhat affecting the project design and feasibility, primarily serves to assist the CM firm in packaging the construction work to yield competitive contractor bid packages. It is necessary for the preparer of a parameter estimate to have more defined data available than at the budgeting phase of a project. It follows that the parameter estimate should therefore be more accurate than a budget estimate.

Scheduling of Design and Preconstruction

Scheduling is a topic that is often emphasized when discussing the construction phase of a project. However, scheduling concepts and techniques, including the use of bar-chart scheduling or the critical-path method, are equally applicable to the preconstruction phase of a project.

The ability to complete a proposed project in a minimal amount of time is as dependent on the timely occurrence of preconstruction decisions and procedures, such as completion of design phases, as the completion of elements of the project's construction. The scheduling and performance of loan approvals, land acquisition, permits, and contract documents affect the start of construction as well as the total project time.

The preparation and maintenance of a schedule for various project design phases is especially critical if the project's construction is to be phased. Phased or fast-track construction is only successful in shortening a project's total time if the drawings and specifications are prepared on a schedule compatible with letting construction packages. The architect/engineer design team should not be delinquent in delivering its services on schedule.

Given the construction project's strong dependence on the performance of the preconstruction as well as construction phases of a project, the CM firm needs to make use of planning and scheduling techniques from the conception of the project through its completion. Planning and scheduling techniques are presented in chapter 8.

Identification of Long Lead Items

Various materials that make up the construction project occasionally become in short supply and are not available for a project without a relatively long lead time. Given the normal process of having the project owner engage various contractors who in turn supply the necessary labor and material, the project owner may in effect inject a delay into the project schedule. The engaged contractor, upon receiving the contract and the go-ahead to start work must order material that may take several weeks or even months to receive. Hence a project delay occurs.

Given the CM firm's knowledge of the unavailability of or long lead time needed to obtain identified materials, the CM firm can advise the project owner to purchase materials when they are set out in the design. The material may not be immediately delivered, but the owner contracts for a delivery date as well as a purchase price. Subsequently, upon the recommendation of the CM firm, the project owner engages a contractor to install or fabricate the material, the delivery of which occurs when the contractor is engaged.

The identification of long lead items can be used by the CM firm to minimize project time and avoid unnecessary delays related to material delivery inefficiencies. The ability to plan long lead items is dependent on the CM firm's knowledge of the existing and projected material markets and the ability to use scheduling techniques that set out necessary purchase and delivery dates. Identifying long lead items is discussed in chapter 9.

Bid Packaging

The determination and setting out of project bid packages is one of the services that can yield the project owner the most benefits. In theory, the CM firm can break out one contract package or a large number. The CM firm's objective is to establish contractor bid packages that accommodate its minimum project time and cost objectives. This means that recognition of the availability of different sizes and types of contractors must be recognized in the bid-packaging decisions. Although each project is unique in regard to the optimal number and scope of bid packages, in practice the number of bid packages often ranges between 10 and 25.

A brief description of the objectives of packaging work was discussed in the previous section of this chapter. Procedures for packaging work, along with examples of bid packages, are presented in chapter 9.

Awarding of Contracts

The CM firm may or may not contract directly with contractors to perform a project's construction work. Usually the contractors in a CM process contract directly with the project owner. However, it is possible that the CM firm will sign a contract itself with the individual contractors.

Independent of whether or not the CM firm contracts directly with the project contractors, the firm does play an important role in awarding contracts. The CM firm makes recommendations to the owner as to which contractors should be invited to bid on various work packages and also advises the owner when and what contracts to award.

The CM firm may occasionally make recommendations to the project owners to award a contract to a contractor other than the low bidder. This recommendation may be based upon the CM firm's appraisal of the ability of the contractor to perform on schedule or to perform quality work. Similarly, the CM firm may recommend that the project owner waive a bonding requirement for selected contractors. This recommendation may be made

Preconstruction Services

based on the CM firm's appraisal of the contractor's ability and experience in performing the work, independent of the contractor's inability to obtain bonding.

The CM firm's objective in awarding contracts, or in recommending to the owner which contracts to award to specific contractors, is to obtain a team of contractors that will deliver minimum project time and cost and also perform quality construction work. In this regard, the CM firm should also perform the service of marketing the owner's project to qualified, competitive contractors in order to ensure that the objectives of time, cost, and quality are attained.

Setting Out Operating Procedures and Responsibilities

The numerous entities and relatively complicated relationships that exist during preconstruction and construction of a project result in a need for well-defined operating procedures and identification of responsibilities throughout the entire process. The need for clear operating procedures and identification of responsibilities is usually recognized by all entities involved in the construction process. Because of too little emphasis on clear setting out of procedures and responsibilities in the preconstruction stage of a project, oversights or unnecessary delays have occurred.

Numerous key decisions and procedures must be executed in the preconstruction phase of a project to ensure subsequent project delay or cost overruns are avoided. For example, projects have on occasion been delayed because of the oversight to obtain a necessary building permit or authorization. On other occasions a designer's oversight to complete a specific phase of the contract documents on schedule has led to a lengthy project delay. The possible use of phased construction in the CM process increases the importance of the design phase being completed on time to enable the construction phases to begin on schedule.

The responsibilities of the entities in the CM process, including the project owner, CM firm, designer, and contractor, are not as defined as they are in the owner-designer-general contractor relationship. This is due to the fact that the CM process is relatively new and also relates to the fact that anytime one adds an additional entity to a process the responsibilities and lines of authority become more complex and less defined.

These potential difficulties can be eliminated if the CM firm draws its attention to them and clearly defines in writing and in policy the operating procedures and responsibilities for the preconstruction phase of a project. The objective of the CM firm in performing this service is to promote communication among all project entities, to enhance the ability of each of the entities to perform assigned and contracted duties, both of which can result in achieving the larger objectives of minimal project time and cost.

Process Paperwork

The execution of a construction project entails the processing of much paperwork. The control of project time, cost, and quality infers the collection of data, the processing of the data, and the distribution and analysis of the processed data.

The CM firm relieves the project owner of much if not all of the paperwork headaches of a construction project. The paperwork consists of material internal to the project, such as the preparation of schedules for design (as well as construction), and external to the project, such as satisfying the needs of external parties. For example, public construction projects require many forms to be processed, or paperwork necessary to obtaining financing from a lending institution may need to be processed.

The efficient execution of paperwork implies the use of accounting procedures and forms. Identification of what data are needed; when they are needed; who is to initiate, process, and analyze the data; and how the processed data are to be filed are just a few of the questions to be addressed in the evolution of an efficient paperwork system.

The CM firm's objectives in executing preconstruction (as well as construction) paperwork are to enhance project controls that reduce or eliminate project time or cost overruns; to enable the project owner to have an ongoing, up-to-date view of the project's status to satisfy the reporting needs of external parties; and perhaps most importantly to enhance timely decision making by all project entities.

CONSTRUCTION SERVICES

The construction services offered by the CM firm, like its preconstruction services, have the objectives of controlling the project's time, cost, and quality. Typical CM services performed during the construction phase are listed in Figure 3.2.

Many of the CM construction services listed in Figure 3.2 are often the implied services of the general contractor in the non-CM process. However, the CM firm is expected to place much more emphasis on the actual performance of these services than the general contractor in the non-CM process. The advocate of the CM process would argue that the failure of the general contractor to perform some of the CM services listed, for example, detailed planning and scheduling, and cost and time control, are in part the reason for the existence and value of the CM process.

Whether or not the general contractor in the non-CM process adequately performs the majority of the services listed in Figure 3.2 is dependent on the general contractor being discussed. Independent of this, it is true that the CM firm's concentration on managing rather than splitting its

Construction Services

> - Detailed planning and scheduling
> - Construction phase estimating
> - Operating procedures
> - Supervision
> - Inspection
> - Testing materials
> - Handling paperwork
> - Handling change orders
> - Cost and time control system
> - Process contractor payments
> - Testing the completed project
> - Marketing the project
> - Property management

Figure 3:2 CM construction services

efforts between managing and building enables it to be in a better position to perform the services adequately.

It should also be noted that some of the CM services listed in Figure 3.2 are clearly not general contractor services. For example, processing contractor payments, testing the completed project, marketing the project, and property management are not general contractor obligations. The objectives of each of the CM construction services listed in Figure 3.2 are described in following paragraphs.

Detailed Planning and Scheduling

As noted previously, perhaps the most important CM service is detailed planning and scheduling. The need for detailed planning and scheduling of the construction phase of a project arises from the complexity of the construction phase of a project. Numerous project materials, labor crafts, contractors, and equipment must be molded into the completed construction project. Any time a production process is complex, the need for detailed planning and scheduling of the component parts increases.

The CM firm's emphasis on detailed project planning and scheduling is in recognition of the failure of contractors to perform these functions adequately. The CM process provides the contractor team leadership in the preparation and monitoring of project planning and scheduling.

Planning and scheduling techniques vary from simple techniques such as a bar chart, to relatively sophisticated scheduling rules using the critical-path method (CPM) as a base. These techniques, along with applications to attain the CM firm's objectives of controlling project time and cost, are discussed in chapter 8.

Construction Phase Estimating

The CM firm's major use of its estimating skills is directed to the preconstruction phase of a project. As noted, the firm has an objective of setting out accurate project scope estimates that enhance an optimal project design. During the construction phase of a project, the CM firm may also need to perform estimating services. For one, a project that is phased requires the CM firm to continually prepare package estimates to evaluate contractor bids. Secondly, independent of the CM firm's input to design, some project change orders are likely to occur. The ability of the CM firm to evaluate the feasibility of any proposed change order, and its ability to negotiate an equitable price for the change order work, is dependent on its ability to estimate the cost of the work accurately.

Unlike the skills or knowledge the CM firm needs to prepare accurate preconstruction estimates, construction phase estimating requires knowledge of detailed costs. For example, to evaluate change order feasibility it may be necessary for the CM firm to have knowledge of the cost of placing concrete for walls, slabs, and so on, or the cost of various finish materials.

The ability of the CM firm to perform construction phase estimating comes from its daily involvement in the construction process. This is the best source of detailed construction cost data. Given this fact, the CM firm that evolves from a contractor background, or one that is also performing construction work in non-CM contracts, is likely best equipped to perform this service. However, this is not to say that the contractor-based firm is necessarily the best CM firm. It merely implies it may be best equipped to perform this particular service.

Operating Procedures

As is true of the preconstruction phase of the project, the existence of clearly defined operating procedures is fundamental to construction phase continuity and the overall objectives of control of time and cost. The identifying and setting out of construction phase operating procedures is a service of the CM firm.

The operating procedures relevant to the construction phase of a project may be expressed orally or in writing to the project entities by the CM firm. Independent of the means of communicating the procedures, the communication should be precise and should clearly establish lines of authority and identify the responsibilities of each entity. The CM firm acts as the leader or orchestrator in the construction phase. Identifying the entity responsible for various project clean-up requirements, identifying the procedures to be followed for project change orders, and setting out the communication process for contractor grievances and claims are just a few of the many operating procedures that must be clearly established.

In addition to having the defined operating procedures serve the objective of delivering the project owner minimum project time and cost, the CM firm's clear identification of operating procedures enables each of the project contractors to realize their profit goals. Indecisive policies or a lack of leadership will result in confusion and uncertainty that ultimately results in an unprofitable project. In effect, by setting out clearly defined operating procedures, the CM firm assists both the project contractors and the CM firm's client.

Supervision

The time, cost, and quality success of a construction project is strongly related to the daily leadership exhibited at the job site. This leadership is provided by the CM firm through its supervision service. Supervision is often viewed as a "policing" role. In regard to the construction process, the supervision function performed by the general contractor in the non-CM process might be viewed as a process whereby the superintendent makes sure all labor and subcontractors are working, material is on the job site on time, and so on. However, supervision should be more than policing or checking-up on individuals. Supervision can and should be a leadership function.

The CM firm's on-site project manager or superintendent is the means of providing project coordination and communication to the on-site contractors. The implementation and monitoring of project schedules, the revision of project estimates, the execution and processing of change orders, and the execution and interpretation of the previously discussed operating procedures are part of the CM firm's supervision function.

In order to effectively supervise the construction phase of a project, it is necessary that the CM firm work with the contractors rather than merely tell the contractors what to do. The most effective supervision promotes an atmosphere where no supervision is needed. This doesn't imply that supervision or leadership is unneccessary. It does imply that the best supervision comes from the supervisor gaining the respect of subordinates by means of decisive decision making, demonstrated ability, and the willingness to respect the interests and knowledge of the subordinates.

On most projects, it is necessary that the CM firm provide on-site supervision by means of full-time project managers or superintendents. The cost of these project managers or superintendents (the number required would vary from as few as one for small projects to several for large projects) should be budgeted in the CM firm's fee.

Inspection

In the non-CM process, project inspection is usually the responsibility of the owner's architect/engineer design team. The CM firm normally takes over this function in the CM process. Whereas project supervision is aimed primarily at controlling project time and cost, the inspection function has the objective of ensuring that the construction is performed to the quality set out in the contract document specifications. To perform the project inspection function, it is necessary to have a thorough understanding of the specifications and to have the ability to evaluate the quality of construction. This implies a knowledge of construction materials and methods.

The performance of the inspection function implies certain liabilities for the CM firm independent of its obligations to deliver the project owner acceptable project time and cost. Should the CM firm approve construction work that does not satisfy architect/engineer design specifications, the firm is likely liable for subsequent project inadequacies, even building failure. On the other hand, should the design prepared by the architect/engineer firm lack structural integrity and subsequently fail, the CM firm is likely not liable for the building failure, independent of its inspection function.

If a CM firm also engages to perform construction work on its project, it can be argued that its dual role is no longer consistent with its agency function. In effect, it is being asked to inspect its own construction and the owner loses checks and balances on an important project variable, the project quality. To some degree this is an argument for not allowing the CM firm to perform construction work. Of course, it is also possible to engage the design firm to perform the inspection service even when the CM process is used. Although eliminating the concern over the loss of checks and balances, the CM firm is better equipped than the design firm to perform the inspection function.

Testing Materials

The material testing function is similar to the inspection function in purpose. Both functions have the objective of ensuring construction quality. However, unlike the inspection function, the testing of materials implies the availability and knowledge of using testing equipment, some of which may be quite sophisticated. Testing machines or procedures include various concrete-testing devices, earthwork sample tests, and tests performed on steel connections.

In the non-CM process, the architect/engineer usually has the ultimate responsibility for testing the in-place materials. However, it may delegate this responsibility to a specialized company, such as a soil-testing company.

Since the architect/engineer design firm does not perform the inspection function in the CM process, it follows that the testing of materials is also best performed by another firm. The CM firm takes on this function and the inspection function. If the CM firm has the resources, both testing machines and expertise, it performs the testing of materials. If not, the CM firm engages a construction testing service to assist in or perform the testing for the owner.

Handling Paperwork

All too often the construction phase of a project proceeds with less than adequate documentation of owner-requested information, external party information, and events that may need documentation to support or defend against possible legal action. The lack of adequate documentation is in part due to the contractor's preoccupation with performing actual construction work and the design firm's passive role in the construction phase of a project.

The CM firm, acting in its management role, is responsible for handling construction phase paperwork. It may delegate some of this responsibility to individual contractors. However, its objective in delegating paperwork to contractors is primarily aimed at having the contractors prepare paperwork that supports that being prepared by the CM firm. For example, the CM firm in preparing weekly or monthly reports aimed at preparing revised project schedules or estimates may require individual contractors to submit progress reports on a timely basis.

Documentation of daily events, including work performed by contractors, men at the job site, material delivered to the job site, and usage of equipment, is a necessary part of the construction phase in order to minimize potential legal disputes and to prevent misunderstandings. Written documentation is the best evidence in regard to project control, support of accounting records, and potential legal disputes.

Handling paperwork also implies an efficient distribution of paperwork to relevant individuals and proper filing. Much of the paperwork performed by the CM firm is for potential use should some uncertain event occur. An efficient paperwork flow, including a filing system, is fundamental to the use of previously processed data. The ability of the CM firm to handle paperwork during the construction phase of a project is dependent on its use of accounting-related skills and procedures.

Handling Change Orders

Change orders to a project often prove costly because the owner is at the mercy of the on-site contractor, and the owner's ability to negotiate the work at a favorable price is diminished. Change orders are also looked upon unfavorably by the project owner because they result in the project cost, and perhaps time, being different from what was first conceived or budgeted.

The reasons for project change orders can be traced to different causes. Some change orders relate to design omissions, others to the owner's change of preference, a project scope change, or because of a totally unexpected construction event. The CM firm has two objectives in handling change orders. For one, the CM firm attempts to reduce the number of change orders in the design phase. Its introduction of construction knowledge into the design process should reduce owner or design omissions as well as subsequent change orders stemming from the unavailability of materials.

Independent of the use of the CM process, the existence of some change orders during the construction phase of a project is inevitable. For these change orders, the CM firm's objective centers around the efficient and economical processing of it. This implies adequate documentation as to the cause of the change order, the estimation of the cost, and the terms of the subsequent performance of the work by an engaged contractor.

The efficient and economical processing of a change order is also dependent on the ability of the CM firm to negotiate an equitable price for the work with a contractor, usually an on-site contractor. The advocate of the CM process argues that the CM firm's everyday involvement at the job site enables it to have a better working relationship with the project contractors than the project designer. This improved working relationship puts the CM firm in a negotiating position with the contractors.

Cost and Time Control System

An important element of control is the comparing of actual performance to planned performance. This element of control is necessary to controlling both construction phase time and cost. The project estimate and the previously discussed project plan and schedule represent the planned performance of the construction project. The project estimate, plan, and schedule serve as the basis for subsequent project control. As noted earlier in this section, the CM firm plays an active role in the preparation of both the project estimate and the plan and schedule.

Events occur throughout the construction phase of a project that result in the project's time and cost deviating from the planned performance. Delays in obtaining project materials, weather, and contractor inefficiencies are just a few of the events that commonly occur during the construction phase of a project. The CM firm's ability to control or correct these potential time and cost problems is dependent on its ability to know that the problem is occurring. Because of the complexity of the construction project and because it is impossible for the CM firm to stand over each construction activity at all times, the ability to detect potential project inefficiencies is dependent on the implementation of a data collection and processing system. Included in this system are the use of several forms, including daily reports, job progress reports, contractor progress reports and claims, and external conditions such as weather.

A control system for construction phase time and cost must include the following three features. It must

Construction Services

   1. Be timely
   2. Compare actual versus planned occurrences
   3. Analyze and correct the problems

It is obvious that the potential to control a project's time and cost depends on the timeliness of the CM firm's effort. The sooner a potential problem is identified, the better the chance to correct it. Depending on the scope and duration of a project, the CM firm's comparison of actual versus plan by means of data collection must be performed weekly or monthly at a minimum. A daily control system would be ideal. However, the inability to process data daily may prohibit such an analysis of project time and cost.

Once a potential project time or cost problem is detected, it is necessary for the CM firm to analyze and correct the problem. Only then is the control function complete. Changes in the layout of the construction work, earlier procurement of material, or adjusting the defined operating procedures are examples of actions that may correct an evolving time and cost problem.

In some circumstances correction of a construction process may not be the cure. The project overrun may stem from an inaccurate project estimate or plan. However, without a control system the determination of the cause of a problem and its correction are not possible.

Process Contractor Payments

Typically, the construction team obtains payment from the project owner by means of progress billings. These billings represent compensation for the work performed. For example, if a contractor is 25 percent finished with the defined contract work, in theory it has a right to bill the project owner for 25 percent of the total contract amount.

The project owner is often far removed from daily involvement in the work of the contractor. Also, the owner likely has little detailed knowledge of construction costs. The result is that the owner could potentially be overbilled and subsequently overpay a contractor. In addition to losing the time value of its overpayment, the owner's overpayment can prove troublesome should the contractor fail financially before the project completion. In all likelihood an owner's overpayment, along with the contractor's failure, will result in the project owner having little or no recourse for the overpaid amount.

The CM firm's involvement in the construction of the project and in project estimating place it in a position of being able to evaluate the fairness of contractor-requested payments. The CM firm approves contractor payment requests to the client, with the objective of protecting the owner's assets.

Some contractor overbillings are an accepted practice in the construction industry. The theory for their acceptance is that the contractor needs to overbill the owner to offset the fact that the owner may hold back a retainer (sometimes 10 percent) from the contractor and the fact that there is usually a time lag between the time the contractor makes construction expenditures and subsequently receives payment.

Although contractor overbillings may be common in the industry, the fact remains that two wrongs don't make a right. The fact that a project owner holds back payment due to a contractor, and the contractor in turn overbills the owner, results in unnecessary owner and contractor costs. The CM firm should try to discourage contractor overbilling by taking steps to have the project owner make payments that are on time and are equitable to the contractor. By having the project owner pay the contractor on time, the CM firm helps the contractor, with the potential of helping the owner by not strapping the contractor for cash, which results in performance delays.

Testing the Completed Project

Once a project is near or at completion, there needs to be a start-up of testing of the functional elements of the project. Although each of the project contractors is responsible for the condition of its in-place work, several contractors may be already off the job-site when the overall project is completed and the owner is ready for occupancy. For example, the plumbing contractor may have performed its work, had it tested and approved by the CM firm, and subsequently left the job-site before the completion of the project.

Included in the possible services of the CM firm is the final testing or approval of all project elements, components, and subsystems for the owner. Should any of the previously approved elements, components, or subsystems prove inoperative or inadequate, the CM firm acts as the owner's agent in pursuing the responsible contractor to repair the problem. Absent of a general contractor, the CM firm serves the project owner in this testing or final approval of the project.

It is also possible that certain end-of-construction-phase work may be difficult to delegate or contract to individual contractors, for example, final project clean-up work, and the CM firm may take responsibility for performing this project wrap-up work for the owner. Obviously the CM firm must include the cost of this work in its fee.

Marketing the Project

Construction projects are built by varying types of owners for varying purposes. In general, projects are built for one of the following three purposes:

1. To accommodate the owner's business objectives
2. To provide the owner a convenience, such as shelter or a transportation system
3. For potential investment income for the owner

Projects of the first two types are used by the project owner after the completion of the project. The CM firm's services for these types of projects normally terminate with the completion of the construction phase of the project or with the CM firm's testing of the completed project.

Projects of the third type are *investment projects*. The owner builds these projects with the intent of renting or leasing the space. Apartment buildings, office buildings, and shopping malls are examples of these types of projects.

Given the appropriate skills, a CM firm may extend its services beyond the construction phase of a project to assist the project owner in securing tenants. The CM firm may also act as the owner's or tenant's agent in administering contracts and monitoring construction of any tenant-required finish work to the rented or leased space. The CM firm's involvement in monitoring the design and construction phases of the owner's project may enhance the firm's ability to perform the owner's marketing of custom finish work.

The skills required to market the project to tenants may be removed from the typical CM service. Much of these skills relate to selling real property, which may not be one of the CM firm's specialties. As is true of all potential CM services, the firm without the skills necessary to perform a specific service should not market or attempt to perform the service. On the other hand, the firm equipped to provide the service can offer the project owner a more complete package of services and thereby better serve the owner.

Property Management

As was true of the above service, property management might be considered as a fringe CM service. Property management refers to the performance or administering of the ongoing maintenance and/or repair of a construction project after the completion of the construction phase.

As projects increase in complexity, the cost and skills necessary to maintain the project increase. The maintenance and/or repair of complicated electrical or mechanical systems may be beyond the ability or knowledge of the owner. Although the owner of a manufacturing facility may employ an in-house staff to maintain or repair its project, the owner of investment projects will often seek this help from external parties.

A property manager does not necessarily maintain an owner's project with its own staff. The property manager may serve as an administrator or manager of the property. It may hire janitorial, mechanical, or service-oriented firms to perform the maintenance or repair of the project.

Given its administrative or managerial skills, and its acquaintance with the owner's project due to its involvement in the design and construction, the CM firm may be able to perform the property manager function effectively for the project owner. Its objective in performing this service is to control the project owner's ongoing project cost as well as the project's initial cost.

Although property management is a possible CM service, many CM firms may not offer it due to their interest in concentrating their skills and resources on the preconstruction and construction phases of a project. Property management, unlike preconstruction or construction services, is an ongoing service that may distract from the ability of the CM firm to perform its primary services.

LIABILITIES OF THE CM FIRM

The liabilities of the CM firm are dictated by its agreement or nonagreement to a guaranteed maximum cost for the project. Obviously, the CM firm signing an agreement that contains a guaranteed maximum provision is liable for project costs in excess of the guaranteed amount. The CM firm signing an agreement without a guaranteed maximum provision, sometimes referred to as a *professional service* CM contract, is without this ceiling cost liability.

The fact that a CM firm does not sign a guaranteed maximum contract does not infer that the firm is free of all other liabilities. Any practitioner, be it a professional service or a production function, is liable for its actions and decisions or lack of them. If a CM firm makes a decision leading to a subsequent project cost overrun or contractor claim, and the decision is judged to be one that a "reasonable" man would not expect from a CM firm in carrying out its contract duties, the CM firm will likely be judged liable for damages.

The key to the determination of a CM firm's liabilities (other than those inferred by a guaranteed maximum contract provision) is based on the *reasonable man concept*. That is, if the CM firm performs (or does not perform) an action or decision that is judged inappropriate (or incorrect) as to what a reasonable man would judge appropriate, then the CM firm is liable. On the other hand, if a CM firm acts as a reasonable man would expect a CM firm to perform, independent of a subsequent event resulting from the CM firm's actions, the CM firm is likely free of any liability. For example, assume a CM firm recommends during the design phase that a certain type of masonry block be used and the architect designer agrees. Subsequently the blocks are determined not to be available, which results in a lengthy project delay and related increased costs. If the blocks were not available at the time the CM firm recommended their use, and in the eye of a reasonable man the CM should have known this, the CM firm is likely liable for the costs resulting from the delay. However, if the blocks become unavailable due to a strike subsequent to the CM firm's recommending them, and it is judged impossible or unlikely that the CM firm could have forecast this, the CM firm is likely free of liability resulting from the delay.

Liabilities of the CM Firm

The CM firm is not judged liable for the integrity of the architect/engineer's design. For example, if a concrete slab fails because it is too thin, the CM firm is likely not liable, independent of its approval or disapproval of the prepared design. However, to ensure that it is not liable for the inferior design, the CM firm has to be careful to stress that it is an adviser of the availability and cost of material in the design phase and not responsible for the final selection of material or the structural soundness of the design.

If the CM firm contracts with the project owner for construction phase inspection and supervision services, it is liable for damages resulting from contractor nonadherence to the contract drawings or specifications. For example, if an on-site contractor places five inches of thickness of concrete in a slab specified to be six inches by the designer, and the CM firm approves the construction, the CM firm is likely liable for any damages relating to the inferior construction. Naturally the contractor is also liable. The project owner will probably seek damages against the entity it judges most responsible, but also most likely able to pay.

The CM firm's responsibility for contractor performance is dictated by its contract agreement with the contractors and the owner. If the CM firm contracts directly with project contractors (which is not typical in the CM process, especially for small- or medium-sized projects) then the CM firm is in effect acting as a general contractor and is directly liable to the project owner for contractor performance.

Even without a contractual relationship with individual project contractors, the CM firm may be judged liable for their performance if in fact the CM firm acts as a supervisor and is held to be directing the contractors. This is why some CM firms go out of their way to express both orally and in the owner/CM agreement that they *observe* rather than supervise the construction phase of the project. However, even when the CM firm contracts to observe rather than supervise, it is likely going to be held responsible for its actions or decisions under the reasonable man principle of law.

Many of the CM services discussed in this chapter are not easily quantified or taught. To some degree, services like setting out operating procedures and processing change orders involve procedures that must be custom designed to a specific project. Although some CM firm services cannot be easily quantified or set on paper, others require the use of specific management tools and techniques. Several of the more commonly required management tools or techniques that result in effective CM are presented in the following chapters. These tools enable the performance of the key CM services discussed in this chapter.

EXERCISE 3.1

STANDARD CM CONTRACTS

Appendixes A and B illustrate standard forms of contracts for a CM firm and project owner as proposed by the AIA and AGC. Although the agreements have similarities, they differ in regard to certain considerations. Review both agreements in the appendixes and contrast them in regard to CM services provided and CM firm liability.

EXERCISE 3.2

CM SERVICES

The chapter indicated that one of the most important CM services provided by a firm is the identifying of a potential project owner's needs. These needs are dependent on the type of project being considered and the in-house skills of the potential project owner. Given the following type of project owner and proposed project, rate the listed services in the order that you believe most important regarding the owner's project time, cost, and quality objectives for each condition stated. List them in the order of 1 being most important, 2 next most important, and so on.

Exercise

		School Board Grammar school	Group of Doctor Investors Office building	Executives Manufacturing plant
a.	Secure financing			
b.	Tax analysis of building			
c.	Value engineering			
d.	Phasing contracts			
e.	Parameter estimating			
f.	Owner purchased material			
g.	Project planning and scheduling			
h.	Project supervision			
i.	Property management			
j.	Market project			

4

Marketing CM and Selecting a CM Firm

CONSTRUCTION MANAGEMENT: A PROFESSIONAL SERVICE

Construction management is a professional service; therefore, the CM firm must be conscious of how to market professional services if it is to be successful. A firm can possess all of the necessary skills, experience, and services of a good CM firm; however, if the firm cannot convince a potential project owner of the benefits of the CM process, or be able to sell itself as the firm to be engaged, its skills, experiences, and services lay dormant.

Marketing CM services is quite different from marketing general contracting. The marketing tool of the general contractor is its ability to be low bidder in a competitive bidding process. In effect, the general contractor is somewhat removed from having to sell or promote itself. However, the CM firm is seldom hired based on its low bid. Although the CM firm's fee is considered by the project owner, the owner places emphasis on qualifications and experience in selecting its CM firm. In order to make a potential project owner aware of one's qualifications and experience, the CM firm must market itself. This involves the use of personal contact, promotional material, and selling concepts.

One of the first steps in a successful marketing program is the identification of potential clients. Undoubtedly, some types of projects are more suited to the CM process than others. Types of projects and owners that lend themselves to the CM process are discussed in this chapter.

Although the CM firm must be attentive to its marketing ability, the potential project owner must be equally attentive to its selection of a CM firm. The engagement of professional services entails an evaluation of the company's ability to perform. Too often a project owner has put too much emphasis on the potential CM firm's fee in its selection rather than on qualifications, skills, and experience. This is undoubtedly due in part to the fact that the owner can easily measure and compare competing CM firm's fees. However, evaluating qualifications, skills, and experience is more difficult. As a first step in evaluating a CM firm, a potential project owner should identify the qualifications and skills it judges appropriate for its CM firm. Recommended qualifications and skills are discussed in this chapter.

TYPES OF CM PROJECTS

At least three variables determine the types of projects compatible with the CM process. The larger the dollar value of the project, the more complex its design and construction, and the fewer the skills possessed by the project owner, the more compatible a project is with the CM process. Therefore, each project must be evaluated on its own merit regarding its eligibility or feasibility for the CM process.

Independent of individual project evaluation, certain types of projects tend to be more compatible with the CM process than others. The following are often worthy of consideration.

Industrial and Commercial Projects

Industrial and commercial projects are usually of a large dollar magnitude and complex in both their design and construction. Projects such as sewage treatment plants, large office buildings, and shopping complexes involve the expenditure of millions of dollars and require the meshing of a number of design skills, labor crafts, materials, and contractors for their completion.

Although industrial and commercial projects qualify for CM consideration, one must recognize that this market is constrained because industrial or commercial clients often either maintain a competent in-house engineering and construction staff or engage design-constructors to undertake large-scale projects and facilities encompassing construction management.

Industry does offer several different market opportunities that warrant consideration of the CM process.

> The medium-size, regionally based manufacturer or commercial enterprise with a once-every-ten-years building program can find the full spectrum of CM services invaluable. Assistance during the planning and design phases, the ability to effect an accelerated phased construction schedule, and an owner-implemented cost-control system are only three of the services most attractive to this client. This market can be approached readily and directly through both the plant engineer as well as the chief executive officer.

> A second opportunity is to augment the management needs of the large industrial or commercial client that has a competent in-house engineer-

ing and construction staff. Examples of these requirements might include designing and maintaining a critical path method (CPM) project schedule, undertaking construction cost auditing and surveillance, project design review and value engineering, or simply providing an on-site supervisor. Another example could be to develop careful performance specifications and contract documents for the procurement of design-constructor services. These services may be meeting client overload conditions, or they may be helpful to upgrade or augment in-house design and construction capabilities.

Another opportunity is to provide certain "nondirect" CM services. Examples might include a construction counseling service for senior management to identify and define an impending construction decision; to explain construction options; and to assist in organizing a construction enterprise. Other nondirect services might include offering construction management training and management development seminars, and offering standard software packages, for example, scheduling or cost-control systems.

Schools, Universities, and Hospitals

Institutions are finding the full spectrum of CM services to be increasingly valuable. Because few of these institutions maintain in-house design and construction staffs, they are in need of competent construction services. Certain complex scheduling needs can only be met through the skills of a knowledgeable CM firm. Frequently, project cost auditing and surveillance are specialized services best undertaken by a qualified CM firm.

Boards of trustees, faced with rigorous fiduciary responsibilities, are becoming increasingly held accountable for their decision-making authority. They are stewards of the institution, not entrepreneurs, and thus are not risk takers. Boards need the skills to make prudent decisions from the inception of the construction program. Usually lacking these construction skills, it is essential that competent (nonadversary) construction management be secured for the board's use and counsel. An experienced independent CM firm can assure a board of trustees that it has not merely accepted the lowest bid (often a naive measure of cost effectiveness), but that it has fully met all of its fiduciary responsibilities.

An increasing number of schools, universities, and hospitals are recognizing that the full-spectrum CM firm offers the management tools that assure the swiftest on-time facility startup consistent with reasonable budget constraints, coupled with the assurance of carefully defined standards of design and construction quality.

Banks and Financial Institutions

With substantial financial commitments at risk, banks and financial institutions are the third party in almost every construction venture. Although often hesitant to play a directing role in design and construction, they are most concerned with the proper management of the construction project to assure on-time startup within an agreed-upon cost envelope. Disappointment with various real-estate developers, architects, and general contractors makes many lending institutions responsive to the introduction of an accomplished CM firm to assure the proper development and completion of a construction project. These third parties are also especially interested in diligent on-site construction cost auditing and surveillance, and in the installation and maintenance of sound cost-reporting and control systems.

Public Sector

Several governmental agencies have been the leading proponents of construction management. In their view, construction management not only rationalizes the entire construction process, but it provides the owner an interdisciplinary approach to handling large and complex projects. It pulls together a fragmented industry and allows architect-engineer, construction manager, urban planner, and a variety of specialists to work together as a project team.

The CM process has also been adopted by other federal authorities. Several federal agencies are striving to get the CM process accepted by municipalities for all major waste-water treatment facilities to speed the projects and cut their costs.

Although some agencies recognize the benefits of using independent CM firm services, the presence of a large tenured staff in the agencies suggests that it may take some time before they are receptive to engaging construction managers for large-scale projects. However, several of these agencies do recognize the desirability of construction management for some of their larger projects.

Utilities

Public utilities inevitably maintain sizable experienced engineering and construction staffs that can assume many construction-management tasks. Further, their primary needs (generating stations) are customarily undertaken by specialized design-constructors. Finally, a 200 to 300 man construction force is common to meet peak maintenance requirements during outages as well as to undertake modifications of existing facilities. It would appear that the public utility has little need for the services of the CM firm.

Although public utilities may not be a primary market, there are areas where the CM firm can be effectively engaged. Construction cost auditing and surveillance are areas of increasing concern to public utilities' managements. Construction labor relations expertise is of increasing interest to assure the utility of its full rights and compliance with the labor contract for each trade. Another area of interest is developing careful performance specifications and the contract documents for the procurement of design-constructor services. Field and shop inspection to assure equipment and construction quality standards present other ways in which supplemental construction management services can be engaged profitably by the public utility.

Interior Design and Tenant Improvement

Interior design, tenant improvement, and renovation work customarily introduce procurement, coordination, and scheduling problems that effect the dollar value of the construction. The tough management tasks presented by this kind of project are often not done or are done very poorly; thus, this work is frequently characterized by cost overruns, coordination breakdowns, and annoying (and costly) scheduling delays. Interior design, tenant improvement, and renovation work present a situation where the skills of a professional construction manager may be critical to successful project completion.

An interior design firm or an architect concentrating on this kind of work will welcome the complementary resources of the skilled CM firm. Similarly, construction/installation work is usually undertaken by smaller and often specialized subcontractors who may also welcome the presence of a strong construction manager. The fragmented nature of this market may make it difficult to reach the key decision maker, but an understood need for competent management of interior design, tenant improvement, and renovation work suggests that this can result in the advantageous application of CM process.

Investment Property

Investment property, including office buildings and apartment complexes, is often built by an investor or group of investors with no knowledge of design or construction. For example, the potential owners of a speculative office-building complex may be professional individuals like doctors, lawyers, and accountants. Often these investors are primarily interested in a project as a tax shelter.

Such potential project owners seldom have the time, knowledge, or desire to get involved in the day-to-day process of orchestrating the design or construction of their projects. Instead, they often prefer to turn over total responsibility to an agent. The CM firm can serve the potential owner or owners of this type of project by serving as the owner's agent from project conception through construction. Also, if after-construction services such as project marketing and property management are desired, the CM firm can include these services.

In effect, investment projects offer the CM firm the most capacity to perform its full set of services without owner indecision or constraints. Often the potential owner of investment property places the CM firm in a position of a "true" agent with few if any constraints. Given this, a CM firm offering a broad package of services has the potential to yield the owner a very high cost-benefit ratio service.

MARKETING THE CM FIRM

Finding work through promotion of the CM firm and its services is the key to success. Unfortunately, the technical skills of a firm are wasted unless the selling or promoting function is accomplished. Numerous books have been written and theories developed as to how to sell. The fact remains that selling can only in part be learned. Personalities, motivation, and verbal skills, although difficult to learn, are vital to selling one's product. Selling one's services starts with searching out potential clients. To the CM firm this means finding potential project owners. How the CM firm becomes aware of projects is referred to as marketing or promoting.

As is true in most professions, several construction-related associations in which a CM firm can apply for membership are readily available. Although the objectives of the associations vary from one to the next, many serve as clearing-houses for the disclosure of upcoming projects to their membership. This service alone can often justify the membership dues.

Subscription to construction-related services can also aid the CM firm in finding projects. Although services such as the Dodge Service concentrate on providing subscribers with cost data and management services, information relating to upcoming projects in various geographic regions is also made available to the subscribing firm. The benefits of any one service must be weighed against its costs. The point is that while a service may help a firm to find work, the cost of the service should not result in such a high overhead that it prevents a reasonable return on the projects the firm obtains.

Many times the least costly and most effective means of searching out work are through attention to announcements and personal contacts. Municipal, state, and federal construction projects are openly advertised in newspapers and bulletins. State boards, such as the Illinois Capital Development Board, notify registered firms of all upcoming projects within the jurisdiction of the board.

As is true in obtaining work in any business, especially in service-oriented businesses, personal contact is the key to finding projects. Contacts through country clubs, church organizations, and charitable organizations can lead to a contract for work. The key to securing work through personal contact is the follow-up. Many times a CM firm has lost a potential project because of not following up a lead. As a result of the firm's failure to develop its initial contact, the owner may be approached by another firm or be influenced by others in evaluating the feasibility of the project.

Unlike several service-oriented businesses, the code of ethics of the construction industry does not prevent the CM firm from openly advertising for work. Yellow Page advertisements are the most common form. However, several firms place a regular ad in a local newspaper, church-related publication, or local news magazine. Others advertise in more widely circulated technical magazines such as *Engineering News Record* or a nation-wide newspaper such as the *Wall Street Journal*.

Some CM firms sponsor local sport teams, such as a Little League baseball team or a bowling team. Although this practice may stem merely from the firm's interest in the sport or the participants, in many cases the motivation is advertising. It is difficult to measure the dollar benefit associated with having the name of the sponsoring firm on the team uniforms, but surely such publicity contributes to the marketing of the firm.

The benefits derived from advertising vary with the work the CM firm performs. Local advertising is effective for small local projects. The returns for advertising are probably the least for the large project. Although advertising gained through sponsorship of a team may favorably influence this type of potential project owner, it is unlikely that the advertising is the reason the owner is drawn to the firm. The pendulum swings back to relatively large benefits from advertising when one considers the large CM firm. These firms often operate in a large geographic area. In fact, the larger firm may take on projects anywhere within the country or even overseas. When this is the case, advertising provides the communication link between the firm and the distant potential project owner. The type of communication needed results in a different type of advertising versus that used by the small firm. In particular, the large firm seeking work in distant locations will get the largest returns from advertisements in publications distributed to these locations.

Similar to the use of advertisements for finding clients is the use of direct mail. In fact, direct mail can be viewed as a form of advertising. The only difference is that direct mail represents a concentrated effort aimed at known individuals or firms, whereas the type of advertising discussed previously was aimed at the general public. Obviously direct mail proves advantageous when the recipient is actually contemplating a project. The fact of the matter is that an individual or firm will seldom initiate a project due to the direct mail of a CM firm. There are exceptions. Nonetheless, direct mail can prove an economical means of having one's firm added to the list of those considered when an owner's future construction plans become reality.

Other than personal contacts, direct mail provides the firm with the most personal means of communication with future clients. However, direct mail communication has to be well worded and aimed at the needs of the client. A poorly written form letter may not only prove impersonal, but it will likely be discarded as another piece of time-consuming mail.

There is a degree of risk associated with direct mail. Its use may lead the recipient to believe the sending firm is hard pressed for work because of unfavorable circumstances or other reasons. Therefore, direct mail advertising has to be well written and well directed if it is to result in benefits.

A firm's success has a way of generating good publicity. If this is recognized, and the publicity is properly used, it can become part of the firm's marketing program. Good publicity can develop from several facets of the CM firm's operations. The building of a unique structure or the use of new construction techniques can aid the firm, as can completing a project well before schedule (especially noteworthy when it is a project that has public attention such as a recreational facility or highway), the receiving of professional awards by individual firm members or the entire firm, the election of members of the firm as officers in professional organizations, and the firm's contribution to community or charitable projects.

The firm should not, however inadvertently, hinder favorable publicity. Regardless of how humble the individual firm members are, publicity should be as much a part of the CM firm's marketing programs as any bird-dogging practice. Seldom, if ever, will good publicity impede the firm's marketing. And best of all, its free. Free in the sense that its distribution is free. Naturally the effort that causes the publicity is not free.

Publicity outlets may be through the local press, trade publications, and professional magazines. Although it is gratifying to have these outlets come to the firm, there is nothing wrong with initiating the communication. The preparation of news releases is the responsibility of the marketing function within the firm. An effective communication link with the local press can prove to be an inexpensive means of obtaining work through publicity.

Just as good publicity can aid in the finding and securing of work, bad publicity can hinder such efforts. Years of sound business practices and success in building projects can be forgotten by the public through a single news release citing an unfavorable activity of the CM firm. An unfavorable suit stemming from the injury or death of an employee, failure to meet project specifications or a project completion time, and the involvement of company personnel in bid fixing or bribery can ruin a firm. In fact, it is fair to say that a single bad news release has a much more unfavorable impact on the firm's marketing than the favorable impact of a good news release. As is true of many areas of reader interest, bad news seems to get more attention than good news. The fact that a CM firm completes construction of a project before the forecasted time often goes unnoticed. However, the public is up in arms should the project be delayed. Such is the difficulty of obtaining and maintaining marketing benefits from publicity.

Needless to say, the CM firm cannot spend unlimited time and money on each of the means noted for finding construction work. Too much effort and cost defeat the very purpose of doing business. The firm must weigh the expected benefits versus the costs for each and every means considered. Models, such as dynamic programming, have been structured to aid the firm in evaluating its total work-finding efforts. However, the fact remains that it is often difficult to quantify benefits from individual marketing efforts. On the other hand, this difficulty should not prevent the firm from attempting to analyze its efforts. Such an analysis may pinpoint wasted efforts as well as where more effort should be directed.

The promoting or selling aspect of marketing does not terminate with the finding of potential construction project owners. The promoting aspect of marketing is only complete upon the signing of a contract by the CM firm with an owner. The success of the CM firm in securing a project contract is typically dependent on an extended list of owner considerations. That is, the project owner is influenced by factors in addition to cost when deciding on a CM firm. Included in these considerations are the following.

> Prior dealings with the CM firm
> Reputation of the CM firm's ability to perform type of work concerned
> Reputation of the CM firm's ability to perform within budgeted time
> Social acquaintance of owner with member of the CM firm
> Hiring practices of the CM firm (for example, affirmative action program of the firm)
> Recognition of "home location" of the firm (that is, the owner may choose to deal with a local firm)
> The CM firm's willingness to "give and take" with owner in current and future work negotiations
> Skills and qualifications of the CM firm

WRITING A WINNING PROPOSAL

For many small projects, the project owner and CM firm come to an agreement after an initial personal contact. For these types of projects, the project owner in effect locates and secures its CM firm through personal contact. For larger construction projects, especially public construction projects, the potential project owner may solicit proposals from prospective CM firms. It may do this by means of a formal document referred to as a *request for proposal*. An example of an owner's request for proposal is shown in Figure 4.1. Given this type of request, the CM firm's ability to obtain work or market itself is dependent on its ability to write an effective proposal. This ability cannot be readily taught. Some individuals are simply better at expressing themselves in writing than others. However, guidelines in preparing an effective proposal can be followed that will result in an improved chance of a favorable reaction.

Writing a Winning Proposal

> *Request for Proposal*
> *Construction Management Services*
>
> **1.0—General Intent of the Request for Proposal**
>
> This request for proposal (RFP) outlines the nature and scope of the professional services of construction management that have been determined as essential to the construction of the Civic Center complex. The outline of services provided in this RFP is not intended to be all inclusive. Any additional services or refinement of work efforts that will assist in the completion of this project should be added to any proposal.
>
> **2.0—Client–CM Relationship**
>
> The Civic Center Authority, the client, will hire the construction manager as its agent to accomplish the planning, control, direction and evaluation of all work relating to construction during the design and construction phases of this project. The project will be organized into two phases: (1) the design phase with subphases of *design development* and *working drawings;* and (2) the construction phase. During the design phase, the construction management firm will be retained under a professional-service, lump-sum arrangement. During the construction phase, the responsibilities of the construction manager will be increased to include guaranteed scheduled control.
>
> **2.1—CM–Architect Relationship**
>
> CM–architect relationship is one of cooperation and teamwork with both being direct agents of the client.
>
> **3.0—The Overall Management Plan for Design and Construction Phases**
>
> The overall management plan to accomplish all work tasks related to construction, during both the design and construction phases, should be prepared utilizing the outline indicated below.
>
> **3.1—Narratives of Strategy to Accomplish Project**
>
> Indicate overall strategy tailored to accomplish this project, address all major issues relative to project planning, project scheduling, phasing, bid packaging, cost control, working relationships, and any other critical problems or unique opportunities of this project.
>
> **3.2—Project Description and Component Identification**
>
> Describe the project and its components and recommend system of identification for use throughout the entire project.
>
> **3.3—Construction Management Control System**
>
> Indicate the concept and methodology of a construction management control system and detail:
>
>   3.3.1 Narrative Reporting System
>   Meetings
>   Estimates
>   Accounting
>   Supervision
>
>   3.3.2 Master Schedule
>   Major Milestones
>   Design Schedule Coordination
>   CM Services Schedule
>   Pre-Bid Schedules
>   Pre-Purchase Schedules
>   Construction Schedules
>   Occupancy Schedules

Figure 4:1 Example of a CM request for proposal

3.3.3 Cost Control Coordination—Design Phase
Initial Cost Estimate—Detailed Components
Design Development Phase Cost Estimate Details
Cost Estimates During Working Drawing Sub-Phase
(25%—50%—75% estimates)
Pre-Bid Final Cost Estimate

3.3.4 Financial/Cash-Flow Study
Cash-Flow Requirements
Establish Accounting System

3.4—Major Tasks Diagram—Design Phase Services
Provide a diagram indicating major tasks, the sequence of activity, and the important interfacing of work efforts. The following services should be considered:
1. Budget review with architects
2. Local construction industry profile
3. Cost control
4. Value engineering and analysis
5. Life-cycle cost analysis
6. Site development coordination plan
7. Long-lead procurement
8. Contract document coordination
9. Others as required

3.5—Major Tasks—Construction Phase
To the extent possible, indicate major tasks that will be required during the construction phase that must be addressed and resolved. A detailed work plan and staff plan are not expected at this time; however, any major task delineation that can be foreseen will be helpful.

3.6—Organization Chart(s)
The project must be organized to ensure a highly efficient decision-making system, maximum communication, and a clear understanding by each participating agency of its role, function, and responsibility. The organization should recognize the two-phase aspect of the project, specify the duties of the CM staff, and indicate important interrelationships among participants.

4.0—Special Considerations
This section of the RFP has been included to highlight considerations for the construction manager that the selection committee has determined as important. Response to each item, if not covered in the previous sections, is requested.

4.1—Coordination of Site Demolition and Preparation
The construction manager will be responsible to coordinate all demolition bids as directed by the City through the Civic Center Authority. The City is the agent to acquire and prepare the site for construction and must turn over the site to the Authority by March. The construction manager, in coordination with the City and the Civic Center architects, will be responsible for planning and coordination of:

Demolition Schedule
Preparation of bid documents
Utility capping, relocations, temporary facilities
Acquiring necessary technical data:
  Special surveys   Unique demolition requirements
  Soil test borings   Other necessary related work

Figure 4:1 *continued*

4.2—Design Development and Review

The construction manager will familiarize himself thoroughly with the evolving architectural, civil, mechanical, electrical, structural, and special component plans and will be responsible to make any recommendations to the Authority with respect to the construction considerations relating to site, foundations, selection of systems and materials, and cost-reducing alternatives.

4.3—Job-Site Responsibilities of Construction Manager

Indicate the specific services to be provided by the construction manager related to job-site control. The following services are expected to be the responsibility of the construction manager:

> Job-site temporary facilities plan
> Security
> Clean-up
> Weather protection
> Comprehensive safety plan
> Temporary utility plan

4.4—Equal Employment Opportunity

The Civic Center Authority will require compliance with the City Equal Employment Opportunity plan.

4.5—Construction Manager's Inspection Responsibility

The construction manager will be responsible for all inspection and testing during the construction phase of the project.

4.6—Utilization of Small Business and Minority Enterprises

The construction manager will propose a program on how to maximize the utilization of small business and minority enterprises during the construction of the Civic Center Complex.

4.7—Pre-Bid Qualification Program

The construction manager will propose a pre-bid qualification program. The term "lowest responsible bidder" should be established and defined.

4.8—Guaranteed Schedule Responsibility

The Civic Center Authority is interested in a guaranteed construction schedule to ensure that planned opening dates will be met. The construction manager should indicate what is necessary from the construction management viewpoint and propose how this may be accomplished.

4.9—Impact of Laws, Statutes, Tax Regulations

The construction manager will be familiar with all applicable State and local laws relating to construction. Any potential problem areas should be addressed.

5.0—Administrative

5.0.1 The overall management plan should follow the outline of this RFP.

5.0.2 It is strongly recommended that the potential on-site project manager be present during the final selection interview.

5.0.3 No specific indication of fee is being asked for as part of this RFP. The selection committee will recommend to the Authority, in order of preference, two firms. If the Authority approves the recommendation, negotiation for a lump-sum fee for the design phase will commence with

Figure 4:1 *continued*

> the first firm. The alternate firm will not be contacted for negotiations unless the Authority and the first finalist cannot agree upon a lump-sum amount.
>
> 5.0.4 A set of the initial design phase presentation drawings are attached.
>
> 5.0.5 Any questions about this request for proposal should be addressed to Mr. A. David, Special Assistant for Development, City of Notown, Phone: 222-3456
>
> 5.0.6 Provide ten (10) copies of the overall management plan for distribution to the selection committee and Authority members.
>
> 5.0.7 All proposals must be submitted not later than Monday, May 23, 19XA, 12:00 noon. They will be collected by:
>
> Director of Public Works
> Room 307, City Hall
> City of Notown
> Phone: 444-5566

Figure 4:1 *continued*

The primary objective of the CM firm in putting together its proposal is to convince the owner that the CM firm has a developed plan to deliver the owner a quality project within a minimal or acceptable time and cost budget. The ability to perform and a plan to perform are what will sell the potential project owner on the CM firm.

The style of the CM proposal influences the reader as well as what is stated in the proposal. The following four elements of style are suggested for an effective proposal.

1. Put the important information up front
2. Make the report understandable
3. Write with an emphasis on the potential client
4. Accent the positive

The reader of any business document or proposal is looking for key phrases or statements. If these phrases or statements appear in the middle or at the end of a proposal they will be overlooked or will lose their impact. Every reader has a limited attention span. The span may be increased if the reader's attention is gained by phrases or statements in the early part of a document. First impressions are important to a reader. Thus the CM firm should set out its skills and its plan for project execution early in its proposal. Subsequent pages in the proposal can support earlier statements and fill in specifics. The CM firm should make the assumption that the later pages of its proposal may not even be read, let alone have an impact on the potential project owner.

Most potential project owners are not literary critics influenced by the use of highly sophisticated language. This does not mean that the CM firm should use improper grammar and weak writing style. However, the more understandable the CM firm's proposal, the more likely the potential project owner will be comfortable with the intentions of the CM firm. A simple, well-put proposal that sets out specifics in layman's language will usually prove more effective than a proposal containing vague concepts or technical jargon that confuses the potential project owner.

A potential project owner likes to feel that a firm it engages to provide professional services is giving it individualized attention. That is, each client wants to feel that the CM firm's best efforts are directed toward completing its project successfully. Too often the CM firm uses the same proposal over and over for different projects. In other words, there is no custom designing of the proposal to fit a specific client. The project owner can see through this type of "off-the-shelf" proposal. A proposal that states the potential project owner's specific needs, identifies the employees to be used for the proposed project, and includes a preliminary execution plan with specifics centering on the project is likely to be positively received by the project owner.

Project owners are often unaware of some of the problems that can occur during a construction project that can lead to time or cost overruns. The result is that the potential project owner is troubled by statements in the CM firm's proposal that raise possible difficulties. The CM firm's proposal should accent the positive. The potential project owner needs to be reassured that the CM firm will be able to control the project time, cost, and quality successfully. In this regard, a proposal that emphasizes what the CM firm is going to do, not what the CM firm will not do, will prove more effective. As a general rule, the reader of a proposal is trying to find statements of what could go wrong. The CM firm's proposal must convince the potential project owner that things can go right. A suggested order for topics included in a CM proposal is as follows.

1. Letter of transmittal
2. Statement of problem and objective
3. Plan for performance
4. Budget or fee for services performed
5. List of skills and qualifications of CM firm
6. Experience of firm
7. Special skills or management techniques utilized by the firm

The letter of transmittal should include a brief statement of the CM firm's skills and also contain a tentative plan to deliver the owner's project within the set out time, cost, and quality constraints. The letter of transmittal should be more than a welcoming and forwarding address letter. It should give the potential project owner a positive image of the CM firm and should result in the owner's having an interest in reading the following sections of the proposal.

The statement of problem and objective section of the proposal should serve to show that the CM firm understands the potential project owner's problem and needs. In effect, this section is a duplicate of the owner's request for proposal. By repeating the owner's problems or needs in its own words, the owner is assured that the CM understands its project and thus will have more confidence in the following sections of the proposal.

The plan for performance section of the CM proposal is the most important and lengthy section. Included here are a tentative time schedule for the project, a proposed organizational chart to manage each aspect of the CM process, discussion of unusual events or conditions and the CM firm's plans for addressing them, and any other information that will create the feeling that the CM firm is going to attain the project owner's objectives.

The budget should be set out after the CM firm has put forth its plan of action. It is important to convince the potential project owner that the proposed service will be of benefit before the owner becomes too locked into what it is going to cost. Setting out a fee before telling the potential project owner what he is to receive may distract the owner from getting the full impact of potential benefits.

On some projects, especially public construction, the potential project owner may request two separate proposals. One proposal, referred to as the *technical proposal*, includes the CM firm's plan for performance and listing of skills. The second proposal, referred to as the *business proposal*, contains a statement of the proposed fee. The reason for requesting two separate proposals is that the potential project owner wants to evaluate the technical proposal on its own merit without being influenced by the fee. Naturally the fee is considered, but not in a way to distract or contribute to the technical proposal.

The list of skills and qualifications and the setting out of the experience of the CM firm serve the purpose of expanding on earlier statements in the letter of transmittal or plan for performance sections of the proposal. Resumes of key individuals to be used on the proposed project along with a listing of completed projects (pictures of the projects plus important cost or project duration data), and letters from satisfied customers are effective in this section.

The section listing or illustrating the specialized skills or management techniques used by the firm might be considered as the "frosting on the cake" section of the proposal. Illustration of the CM firm's unique planning or scheduling techniques, cost-control system, or computerized management tools might prove to be a positive influence on the potential project owner.

This is especially true since most potential project owners like to think that they have hired a one-of-a-kind CM firm to serve as their agent.

SKILLS OF THE CM FIRM

Perhaps no factor is as important in the appraisal of a potential CM firm as the evaluation of its skills. Experience is easy to evaluate, based on the firm's record of projects completed. However, one could argue that experience is not necessarily a plus when it comes to performing a professional service. A CM firm that has consistently failed to deliver project owners' time and cost should be judged as inadequate, despite their experience. In other words, not all experience is good experience.

To perform CM services, the firm should possess certain fundamental skills. Perhaps no two individuals would agree to the same list of skills possessed by a good CM firm. However, most would concede that the following three broad skills serve as a basis for performing CM services.

1. Management skills
2. Design skills
3. Construction skills

Included in the management skills are the knowledge and use of various management tools, including project planning and scheduling techniques. The need for design skills is necessary because of the CM firm's involvement in the preconstruction phase of the project. Construction skills are a necessary prerequisite to directing and supervising the construction phase of a project.

Several large potential project owners, such as governmental agencies, have developed prequalification guidelines to be used in evaluating prospective CM firms. Included in these guidelines are a specific list of skills prospective CM firms are to be measured against. Figure 4.2 gives the list of CM firm skills as set out in the request-for-proposal document prepared by the General Services Administration (GSA). Several state and federal agencies have adopted the GSA qualification document and accompanying list of skills for evaluating prospective CM firms. The list of skills in Figure 4.2 is for selection of a CM firm for an industrial or commercial construction project. It can be argued that a separate list of skills is necessary for other types of projects.

Inspection of the list indicates the broad abilities expected of the CM firm. These range from accounting and legal knowledge to knowledge of specialized types of construction work such as electrical or mechanical construction.

At first glance, one might expect only a very large firm to have enough resources (financial and personnel) to possess all of the skills listed in Figure 4.2. However, it is not necessary to have a license or years of experience to

> 1. Accounting
> 2. Architecture
> 3. Civil Engineering
> 4. Construction Superintendent
> 5. Contract Law
> 6. Electrical Engineering
> 7. Estimating
> 8. Construction Inspection
> 9. Labor Relations
> 10. Mechanical Engineering
> 11. Safety
> 12. Structural Engineering
> 13. Testing Facilities
> 14. Value Engineering
> 15. Management

Figure 4:2   Skills of a CM firm per GSA

possess some of the skills listed. For example, it is not necessary to have a CPA license to have a good working knowledge of accounting. What is necessary is the ability to design and implement a reporting system that predicts project time and cost and can then be used to monitor actual project time and cost.

The same can be said of the other skills. It is not necessary for the CM firm to be licensed to perform architectural work in order to qualify as a CM firm. However, a good working knowledge of the architectural design process is beneficial if the CM firm is to work with and contribute to the project design team effectively.

The listing of contract law certainly does not imply that the CM firm must have a lawyer on its staff. However, the CM firm's ability to foresee legal disputes before they occur will play a major role in eliminating or reducing cost and time-consuming delays resulting from jurisdictional disputes or failure to set out the responsibilities of the project team members. Much of the ability to foresee these disputes or other problems is dependent on the CM firm's experience with the construction process and the contractual relationships of the various project entities.

Some of the skills listed, such as value engineering, are broad in scope. The amount of formal education or experience required to obtain these skills varies considerably. Although the list of required CM skills should not be considered as authoritative for all projects, it does provide the potential project owner a place to start in objectively selecting the CM firm. Ideally, the project owner should continually revise its list of preferred CM skills as it measures the success or failure of its various firms on several projects. However, since many project owners are one-time builders, they need a standardized list to serve as a guideline in the CM firm selection process.

EXERCISE 4.1

WRITING A WINNING PROPOSAL

Figure 4.1 illustrated a request for a proposal that might be used by a potential project owner in seeking to engage a construction manager. The success of a CM firm in writing a winning proposal is dependent on its ability to set out how the CM firm will organize itself to attain the owner's time, cost, and quality objectives. Assuming a CM firm is writing a proposal to the project described in Figure 4.1, design an organizational structure for a CM firm that will convince the owner that the CM firm has a "game plan" for delivering the project on time and within the cost budget.

EXERCISE 4.2

EVALUATING CM FIRMS

This chapter discussed the fact that a potential project owner should be selective in choosing a construction manager. Specifically, the chapter indicated that the experience and skills of the CM firm should be evaluated as well as its fee and action plan. Develop a CM qualification form to include a step-by-step formula with point assignment and considerations to be weighed that can be used by a school board to evaluate CM firms competing for a $2 million school building.

5

Feasibility Estimates and Tax Analysis

IMPORTANCE OF THE FEASIBILITY ESTIMATE

Construction projects are built for various reasons. Public construction is built to satisfy the physical needs of the public at large. Private construction, such as a commercial building or apartment complex, is often undertaken as an investment to yield the owner future monetary rewards. Yet other projects, such as schools, are undertaken to fulfill social and economic needs.

Regardless of the underlying reason for the initiation of a project, each and every project has measurable benefits and costs. The ability to measure these benefits and costs at an early stage of a project can decrease the potential cost and risk associated with it. The rapidly escalating costs of construction have drawn attention to the need for a comparison of the anticipated benefits and costs of a project at the time of its conception. This comparison of benefits and costs is commonly referred to as the *feasibility estimate*. The construction manager can assist the potential project owner by performing the feasibility estimate.

The project drawings and preliminary plans are usually nonexistent when the feasibility estimate is made. As such, its detail and the means by which it is made differ from a contractor's detailed estimate. In particular, the feasibility estimate places strong emphasis on the projection of benefits because these benefits may be more difficult to forecast than project costs. Tax considerations are also important in a feasibility estimate, whereas the contractor's detailed estimate will commonly address itself only to costs.

The owner's involvement with a project commonly far exceeds that of the construction contractor. Although an owner may sell the project, this may not occur for several years, if at all. Three factors, depreciation, debt service, and life-cycle costs, are major considerations in a feasibility estimate.

The accuracy, completeness, and reliability of a feasibility estimate are dependent on the use of several skills. These skills include estimating, knowledge of the use and expected life of construction materials, accounting and related tax laws, and project design. These skills are compatible with the CM firm skills set out in chapter 3.

Unless all of these skills are used, the feasibility estimate will lack accuracy and completeness. For example, an accountant lacking construction knowledge may fail to use correct initial project costs and annual maintenance and repair costs in preparing a feasibility estimate. The result will be a correct process but invalid results. Similarly, if the accountant is unaware of the taxable lives of various building components, this may result in failure to take advantage of beneficial tax rulings. The constructor, because he lacks accounting and tax knowledge, will not be in a position to do a complete or accurate feasibility estimate. The same is true of an individual limited to design knowledge.

In addition to having all the skills to perform an accurate and complete feasibility estimate, the CM firm can supplement the skills of the potential project owner's accountant. Even if it does not perform the feasibility estimate, the CM firm can add its construction knowledge to the feasibility estimate. The CM firm's involvement in this early phase can provide other benefits. In particular, the ability to control project time, cost, and quality is in part dependent on a single entity's involvement in all project phases. Secondly, the CM firm can gain the confidence of the project owner early in its performance of services. This owner confidence is needed if the CM firm is to successfully provide subsequent services.

This chapter presents the mechanics of the preparation of feasibility estimate. Tax topics such as depreciation, componentized depreciation, investment tax credit, debt service, and capital gains are also discussed. It is intended that this knowledge, along with construction and project design knowledge, will enable an individual to prepare an accurate, complete, and reliable feasibility estimate.

MEASURING PROJECT BENEFIT

The predicting of a project's benefits is fundamental to evaluating its feasibility. This prediction is often subjective, in that it is based on factors that are sensitive to future events. Factors such as future housing demand, industrial production levels, cost of capital, lifestyles, and shifting population trends all relate to the measurement of a project's future benefits. A construction manager may need outside advice in order to predict these factors and measure their importance to the feasibility of a project.

The types of benefits derived depend on the type of construction project in question. Figure 5.1 illustrates a list of types of construction projects and some of the benefits of ownership. A forecast is needed to quantify the benefits obtained from a project. In order to make a reliable forecast, it may be

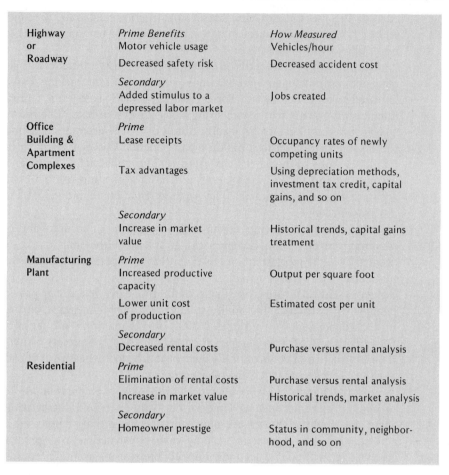

Figure 5:1 Benefits from projects

necessary for the construction manager to hire a consultant with expertise in such forecasts. Consultants that offer these services are real-estate appraisers, marketing analysts, management service departments of certified public accounting firms, management accountants, and real-estate developers.

The approach used by the construction manager or the consultant is referred to as *market research*. Market research includes the study of population trends, family households, industrial development, production demand, real-estate occupancy rates, personal income, and so forth. Marketing research firms typically make a forecast based on correlations among various factors. For example, the market research used by a potential housing developer may relate near-term employment trends in a given location to the need for housing. Such a hypothetical analysis is illustrated in Figure 5.2.

Subject	This year	Next year
Employment	24,000	25,200
Employment participation rate	35.3%	35.8%
Population	68,000	70,382
Household population ratio	97.1%	97.1%
Household population	66,000	38,350
Average household size	3.30	3.29
Households	20,000	20,800
Overall occupancy rate	95.2%	95.4%
Housing units	21,000	21,800

Figure 5:2 Correlation analysis for potential housing developer

To evaluate both a project's benefits and costs on the same basis, the benefits of a project are normally converted to a monetary value per year. This is done by transforming increased production rates or anticipated rental or leasing receipts into an estimated annual cash inflow.

OWNER'S ESTIMATED PROJECT COSTS

Perhaps easier than estimating the project benefits, but no less significant to a project's feasibility, is the recognition of project costs. These costs include initial construction costs, design costs, finance costs, and maintenance and repair costs. Although construction costs and design costs are one-time payments, other costs, such as finance costs and maintenance and repair costs, are paid annually.

The feasibility estimate usually is made using an estimate of initial construction costs. In order to make this estimate the construction manager normally uses cost books or its own cost data. These cost books or other publications give the costs of various types of buildings per square foot or per volume. For example, such a book may indicate a building cost of $45.50 per square foot for an apartment complex. Some of these books also account for various quality building types. For example, they may indicate a cost per square foot for cheap construction, average construction, or high-quality construction. It is wise for the user of these books to determine where their project fits in the defined categories. Needless to say, the construction manager has to be careful using such cost books. The costs per square foot indicated in them are approximate for several different reasons. For one, the costs reported are usually for a rather large geographic area. That is, they are the average costs for numerous buildings of the same type built in different locations. For a building in a specific location, the cost of construction may differ substantially from the average cost reported. The second difficulty with using the average cost in these books is that no two buildings are alike. Even when using the books that give different qualities of construction, it is unlikely that the proposed building will be identical in design, material, and other price factors as those published in the cost books.

Building Costs

The fact that there are difficulties with using cost books does not take away from their use in preparing estimates. These books provide the construction manager with a quick way of determining an approximate cost for a project. These costs are usually accurate enough to be able to determine the project's feasibility. This fact, along with the unavailability of complete sets of drawings and specifications, results in the value of these approximate costs. One solution to the difficulty of using a cost book is for the construction manager to compare the costs published in two or three books. By comparing the costs, the construction manager can be assured that the estimate is accurate. This is especially true if the cost per square foot or cubic foot varies little from one book to another. Three of the cost books available to the construction manager are *The Building Estimator's Reference Book*[1], *The Dodge Manual*[2], and *Mean's Building Construction Cost Data*[3]. Real-estate appraising books also contain such data. In addition, as part of promoting business in a given geographic area, various cities publish cost data for public buildings constructed in their geographic area.

Caution should be taken in estimating construction costs to recognize the time difference between when the estimate is made and when construction will actually take place. Too many owners and construction managers have been surprised that by the time they get to actual construction that the bid for their particular project has come in as much as 50 percent or more higher than their initial feasibility estimate. This may be due to an incorrect estimate or to the time lag between the estimating and letting of the project. The construction process can be broken into three phases, the *feasibility and design phase*, the *bidding and letting phase*, and the *construction phase*. On some projects, especially public construction projects, it is not uncommon for each of these phases to take approximately the same time. For example, a project that takes 6 months construction time may take 6 months in the feasibility and design phase, and 6 months in the bidding and letting phase. During the 6 months between the initial feasibility estimate and the actual letting, material and labor costs may increase as much as 5 or 10 percent. The feasibility estimate must recognize this increase in costs.

---

[1] *The Building Estimator's Reference Book*, Frank R. Walker Company, Chicago, Illinois
[2] *Dodge Manual for Building Construction Pricing and Scheduling*, McGraw-Hill Co., New York, New York
[3] *Building Construction Cost Data*, Robert Snow Means Company, Inc., Duxbury, Mass.

The construction manager must also consider design costs in the feasibility estimate. Almost every construction project requires input from either an architect or an engineer. Normally building construction is much more dependent on the services of an architect, but public works projects such as dams and highways require an engineer. Some projects require the services of both. For example, the overall design of a large office building, including the selection of materials, interior finishes, and overall interior layout may be the responsibility of the architect. The structural frame, however, may be the responsibility of an engineer. Most states require the architect or the engineer to be registered with the state.

The fees charged by the architect or engineer depend on several variables. For one, professional fees are dependent on the degree of service provided. That is, they may only provide design services, such as the preparation of the contract documents to include the drawings and specifications, or they may provide construction supervision and letting services. In the CM process the supervision and letting are typically performed by the construction manager. The professional fees charged by the architect or engineer also depend on the type and the dollar size of the project. For unique projects, such as an elaborate church or high-rise office building, a relatively higher fee may be charged than for a more standard similar dollar value project. In addition, it is common for fees to decrease as a percentage of the cost of a project as the cost of the project increases. That is, whereas the fee for a $1 million project may be 5 percent of the cost of the project, the fee for a $5 million project may be only 3 percent.

Similar to construction and design costs, the cost of land is a one-time charge. It is this type of cost that places a heavy burden on the initial feasibility estimate. High initial land costs are often prohibitive to project feasibility. Land costs over the recent years have typically outpaced the cost of labor and materials. The cost of land relative to the construction cost of a given project depends on several factors, including the following

1. The scarcity of land in the area
2. The acreage purchased
3. The relative demand as a function of time for the land purchased
4. The negotiating power of the seller and buyer

Other factors also affect the cost of the land relative to the construction cost. In general, land cost for any construction project can be estimated to be from 10 to 30 percent of the construction cost for the project. This high land cost is especially prohibitive because land is not a depreciable asset and therefore will not provide a tax savings through depreciation as the construction improvement costs will. Depreciation benefits are discussed in the following sections.

Land costs are more easily determined than construction costs at the time of the feasibility estimate. Land is an observable commodity and therefore it becomes possible to get an appraised value for the land to be purchased. On the other hand, the construction improvements at the time of the feasibility estimate do not exist. As such, these costs are much less traceable and more difficult for the construction manager to determine.

Finance Costs

Finance costs may run as little as 5 percent of the construction costs, but more often they run as high as 10 to 20 percent over the life of the project. That is, each year a project may have finance costs associated with it that equal 10 to 20 percent of the initial construction cost for the project. It is not unusual for the total finance cost of a project to equal or exceed the initial construction cost. Finance costs are annual costs, and this stream of costs must be recognized in the feasibility estimate.

The capital that an owner uses to fund a particular construction project comes from one of two broad classifications of funds. For one, the owner may use his own capital or capital raised through additional stockholders. In this case the capital is referred to as *equity capital*. When using equity capital the project owner does not pay a finance cost to an external party; therefore, it is common for a construction manager or owner to neglect to recognize a finance cost in the feasibility estimate when using equity capital for a project. However, it should be noted that had the project owner not built the project from the funds, the funds could have been invested at some rate of return. This potential rate of return should be charged against the new project in the feasibility estimate. This loss of return is referred to as an *opportunity cost*. Although the opportunity cost is normally less than a finance charge, it can effect the feasibility estimate.

More often than not an owner is dependent on other sources of funds for a project. In this case the funds are referred to as *debt capital*. Debt capital is a loan. When an owner builds a project, two types of loans may be needed. One is called a *construction loan*. A construction loan is needed to fund the project during construction. Sources of construction loans include banks and savings and loan associations, and an increasing source of construction loans, Real Estate Investment Trusts. Real Estate Investment Trusts (REITs) are pools of funds raised by groups of investors that are used to fund construction projects. The group of investors gives the potential project owner this money in return for interest. As is true of most construction loans, Real Estate Investment Trust loans typically have a high interest rate associated with them. Part of the reason for this high interest rate is that should the project owner go bankrupt during the construction, the project in its uncompleted stage would not return a great deal of value to the lender.

In addition to the construction loan, the project owner is commonly in need of a *mortgage* loan. The mortgage loan is used to pay off the construction loan for the project owner. The sources of mortgage loans include savings and loans, insurance companies, bonds, and, to a lesser degree, banks. In obtaining a mortgage loan the project owner puts up the completed project as security. Because the completed project has an identifiable market value should the owner default on the loan, the lender of the mortgage loan is usually willing to provide a lower interest charge than for the construction loan. This lower interest rate reflects a lower degree of risk.

Mortgage loans are usually amortized. This means that the owner pays back the principle in payments that include part of the principle and part of the interest. Initially, the payment includes a large amount of interest and a small amount of principle. Over the years, the principle's portion of the payment increases and the interest charge decreases. Although construction loans are usually written for a period to equal actual construction, mortgage loans typically have a duration of 20 to 40 years.

Another classification of loans that the project owner takes on when building a project is a *conventional* or a *nonconventional* loan. A nonconventional, or sometimes referred to as a *subsidized loan,* typically is tied to a government-related program. For example, FHA and VA loans are nonconventional or subsidized loans. The availability of these loans is dependent on the project owner meeting certain qualifications. For example, income normally cannot exceed some predetermined maximum, and the owner may have to show the unavailability of conventional loans. The conventional loan is merely a loan that is not subsidized. The savings and loan is the usual source of conventional loans.

Included in the services of a construction manager is identifying sources of loans for an owner. Once the sources are identified, the construction manager may serve the project owner by evaluating alternative loans and assisting the owner in securing the loan. The cost of a loan is dependent on the following characteristics.

1. The availability of money in the money market
2. The monetary policy of the government, including the Federal Reserve System's charge to banks for capital
3. The financial soundness of the borrower
4. The negotiating ability of the project owner and/or construction manager
5. The degree of risk the lender associates with the project
6. The lender's forecast of the future availability of money and the cost of such money

Other factors, such as the ability of the owner or construction manager to market the project to lenders, also bear on the cost of finance. The point to be made is that the finance cost associated with equity capital or debt capital can be considerable and can weigh heavily on the feasibility of the project. One satisfying point to the potential project owner is that interest or finance cost is a taxable expense and as such yields certain benefits. These benefits will be discussed in a following section.

Operating Costs

Of all the costs relevant to the feasibility estimate, perhaps the most difficult to estimate are the annual operating expenses of the project. These operating expenses are sometimes referred to as *maintenance and repair costs*. Operating expenses are annual costs and are difficult to estimate because the component costs are hard to identify and because operating expenses increase at varying rates over the life of a project. Unlike the finance costs the operating costs can fluctuate dramatically one year to the next. Components of operating expenses include everyday maintenance and repair costs, such as repairs on heating and ventilating components and repairs on exterior and interior finishes, monthly utility bills, custodian fees, and yearly landscaping costs. Numerous other maintenance and repair costs can be identified. Their identification can be instrumental in determining the annual operating cost for a project.

The difficulty associated with tracing each of the operating costs throughout the life of the project can be resolved in a couple of different manners. First, an attempt can be made to chart these costs over the life of the project based on past trends. This procedure assumes that future cost patterns will follow past cost patterns. Recent experience has shown this assumption to be less than accurate. A second and more acceptable procedure for taking variable operating costs into account is to estimate the first year operating cost and assume that benefits will be increased to cover increases in operating costs. This is the assumption that investors in apartment complexes and office complexes often make. That is, should the operating cost of an apartment complex increase 10 percent in the second year of its life, rental rates will be increased to cover this amount. This makes the estimation of operating costs more accurate. However, an implicit assumption is that the owner will be able to pass on the increased operating expenses. Competitive factors may prohibit the owner from increasing benefits at this rate. Nonetheless, the potential project owner or construction manager does have to recognize the variability of operating expenses over time. Failure to do this will result in underestimation of operating costs and therefore an inaccurate feasibility estimate.

Operating costs for various types of projects are not as well documented as construction costs. Most cost books do not attempt to identify operating costs for a particular type of project. Undoubtedly, the reason for this lies in the uniqueness of the cost structure of each project. In addition, energy and utility costs vary substantially from one part of the country to the next. Finally, operating costs are related to construction costs. A building that uses inferior construction materials will likely have higher operating costs. These higher operating costs would be reflected in increased maintenance and repair costs and in higher utility costs.

Perhaps the best source of operating cost data is the information available from owners of similar projects. In addition, the construction manager should be able to give the owner some assistance in estimating future operating costs based on its experience. Although difficult to estimate, the significance of operating costs to the feasibility estimate cannot be overemphasized. These operating costs can exceed finance costs, and the total operating costs for a project can far exceed the initial construction cost. The significance of operating costs has been highlighted in recent years with significant increases in the price of energy.

Evidence indicates that operating costs are often underestimated. Although some of this underestimation is undoubtedly due to the difficulty of predicting escalating labor, material, and energy costs, part of the fault also lies in overlooking several of the factors that make up the operating cost. Related to the operating cost of a project is the fact that sooner or later a project will approach its total estimated life. If a proposed project clearly has a finite life, then another cost, referred to as *replacement cost*, should be recognized in the feasibility estimate. This is the cost of replacing the project after it has become nonfunctional. This replacement cost will likely exceed the initial construction cost due to increasing costs of construction labor and materials. More often than not an owner or construction manager fails to recognize this replacement cost in the feasibility estimate. This failure can only be justified if the owner or construction manager is looking at a project for its returns over the finite life of the original design. Depreciation accounting is an attempt at setting a reserve for the replacement of a project. However, depreciation only recognizes the construction cost as originally determined for the project. In times of inflation, depreciation does not provide a fund or reserve great enough to replace the project. More will be said about depreciation in following sections.

INCOME TAX RELATED FACTORS

The factors relevant to the feasibility estimate are not limited to the cash benefits and cash costs discussed in the previous sections. In particular, depreciation effects, capital gains considerations, and deductions for operating and interest expenses can all dramatically affect the feasibility estimate. In fact, some construction projects are built primarily for the positive tax benefits.

Depreciation

The Internal Revenue Service recognizes that buildings, being capital assets, yield returns to their owners over long periods of time. Since most buildings are built to produce a profit, the IRS allows depreciation expenses to reduce the taxable income from a project. However, rather than let the project owner deduct the cost of the entire project when built, the IRS requires the project owner to distribute the expense of a project over its estimated life. The process of doing this is referred to as *depreciation*.

Depreciation reduces the cost of a project. In effect, depreciation reduces the taxable income of the project owner. This is especially beneficial if the project owner is in a high income tax bracket. For example, should the project owner be in a 40-percent tax bracket, each dollar of depreciation can reduce the project owner's tax liability by $0.40.

Government guidelines determine the life over which a particular building can be depreciated and the method of depreciation that the project owner can use. Naturally the owner would like to depreciate the project as quickly as possible. This fast depreciation provides the owner the benefits of the time value of money. The sooner benefits are received, the higher the monetary value associated with them.

	Asset Depreciation Period (years)
Factories	45
Garages	45
Apartments	40
Dwellings	45
Office Buildings	45
Warehouses	60
Banks	50
Hotels	40
Office Furniture	10
Computers	6

Figure 5:3   IRS depreciation guidelines

Tables provided in various IRS publications state the minimum number of years a building can be depreciated. Such a table is shown in Figure 5.3. The IRS will allow the project owner to depreciate a project in a number of years greater than the minimum, but it is unlikely a project owner would want to do so.

IRS regulations provide the project owner much flexibility in choosing a depreciation method. Methods such as straight-line depreciation, declining

Income Tax Related Factors

balance, double declining balance, sum of the digits, and others are available to the taxpayer who is depreciating capital assets.

The *straight-line method* is perhaps the easiest to use. The taxpayer merely takes the total cost of its assets minus any salvage value and divides that by the number of periods or years the asset will be in use. The result is the annual depreciation charge that is used in calculating income tax liability. Although it is the easiest to apply, the straight-line method is perhaps the least advantageous in terms of time value of money concepts. Faster depreciation methods are preferred. However, the project owner is limited in its choice of methods. In particular, the IRS restricts the number of depreciation methods that can be used for depreciating buildings. The government indicates that the double-declining-balance method (twice the straight-line rate) and the sum-of-the-digits methods can be used for new residential rental property. The maximum allowance available for nonresidential new property is 1.5 times the straight-line rate. Used residential property is subject to a maximum schedule of 1.25 times the straight-line rate.

The benefits of depreciation are illustrated in Figure 5.4. Depreciation itself is a noncash expense to the project owner. As such, there is no cash outflow. On the other hand, for each dollar of depreciation, the project owner receives a benefit through a reduced tax liability. This reduced tax liability increases as the taxpayer's income tax bracket increases.

Year	Initial Cost	Depreciation	Cash Inflow Depreciation[a]
0	$100,000		
1		$30,000	$12,000
2		21,000	8,400
3		16,333[b]	6,533
4		16,333	6,533
5		16,333	6,533

[a] Assumes 40% income tax bracket
[b] Depreciation method changed to straight line

Figure 5:4 Depreciation benefits

Componentized Depreciation

As noted, the value of money is dependent on time. The sooner a project owner receives cash from depreciation, the more the cash is worth. Two means of speeding up the cash received from depreciation are to use an accelerated depreciation rate, and to use the minimal depreciation life for the asset being depreciated. For example, a depreciation rate of 150 percent is preferable to the straight-line rate for depreciating an office building. If it is acceptable under IRS regulations, a 40-year depreciable life is preferred to a 50-year life.

Another significant means of increasing the cash benefits of depreciation is through the componentized depreciation analysis of a building. Componentized depreciation can yield significant benefits by increasing a project's cash flow during the early years of a building and, therefore, increasing the effective rate of return. On occasion, these benefits can turn an unfavorable building venture into a profitable project. The construction manager can assist the potential project owner in optimizing a proposed project's tax benefits through its construction and estimating skills.

A componentized analysis of a building centers around the fact that several of the components of a building have a life less than the entire building. When depreciating a building as an entire entity, the published IRS regulations specify a minimum depreciation life to be used. These published lives are usually the anticipated life of the building frame or the life of the longest-lived component. Examples of these published lives were shown in Figure 5.3.

The productive or useful lives of various building components are often less than the useful or productive life of the building as a whole. For example, the roof of a building may have a 20- or 25-year design life, various electical fixtures may have a life of 10 years, wall-paper finish in rooms may be expected to be useful for 5 years, and so on. The useful or productive lives of the various components are best determined through construction knowledge and experience.

A building has to be broken into categories of useful or productive life to enable a componentized depreciation analysis. It should be emphasized that these component categories are not subcontracts. Whereas it is easy to determine the dollar amount of contracts awarded to specific contractors, these dollar amounts or contracts will serve little benefit in making a componentized depreciation analysis. For example, an $800,000 contract may be awarded a mechanical contractor for all the mechanical work for the construction of a building. However, within this contract, there are likely many elements or components that have different depreciable lives. In other words, a componentized depreciation analysis requires the support of a componentized cost breakout and estimate, rather than a subcontractor cost breakout and estimate.

The previously defined skills (chapter 4) of the construction manager are compatible with those required to perform a componentized depreciation analysis. Three important steps must be performed if a componentized depreciation analysis is to be defendable to the IRS. All of these steps must be performed prior to the start of the tax-reporting process for a project. Ideally the componentized depreciation analysis should be part of the design func-

Income Tax Related Factors

tion. Because of the relationship of the life (relative to tax implications) and the type of material selected, it is obvious that the choice of materials has tax and therefore economic impact on the overall cost of the building. Therefore the tax impact should be considered in the selection of materials in the design phase of a building.

One of the three important steps of a componentized depreciation analysis is the determination of the useful or productive lives of various components. This is best accomplished by setting out various life categories (such as 5, 10, 15, 20, 30, or 40), for each subcontract or building subsystem and then itemizing each component of the subcontract or subsystem into one of the life categories. The knowledge of construction materials, methods, and elements (electrical generators, furnaces, and so on) is fundamental to this task.

Once components have been identified and categorized, a component estimate must be prepared in which the cost of each component (material and installation) must be determined. This is cost estimating and implies the knowledge of detailed construction costs. Finally, the indirect project costs, including the architectural and engineering costs and indirect contractor costs (overhead) are distributed to the calculated component costs. After this has been performed, the sum of the components costs (each reflecting a share of the indirect costs) should equal the expected total cost of the project.

The IRS's acceptance of a componentized depreciation analysis is dependent on its reliance on the individual that prepared the analysis. The IRS puts more reliance on an analysis prepared by an individual possessing construction knowledge than an analysis "pulled out of the air" that is prepared by a nonconstructor. In effect, the involvement of a construction manager in the preparation of a componentized depreciation analysis may enhance its acceptability. It is important to note that the IRS does not tell the potential project owner or construction manager how to break out or cost the components. This is left to the creativity or skills of the individual preparing the analysis.

To illustrate the benefits of componentization, let us consider a $3 million manufacturing plant. Ignoring an investment tax credit and componentized depreciation analysis, we use a 150 percent declining-balance depreciation method and a building life of 45 years to yield a first-year depreciation expense of $100,000. Given a 40 percent tax bracket, this would yield the owner $40,000 of cash flow in the first year of ownership.

Let us assume a componentized depreciation analysis resulted in the following building breakdown. (See following page.)

Using 150 percent declining-balance depreciation for the non-investment tax credit components and 200 percent declining-balance depreciation for the components eligible for investment tax credit (discussed in the next section), a first-year depreciation charge of $488,500 is calculated. For the 40 percent bracket taxpayer, this yields first-year tax savings or cash flow of $195,400. This difference in the first year cash represents over 6 percent of the anticipated cost of the project. This added cash benefit from componentized depreciation significantly increases the project's rate of return.

Component Life	Amount Eligible	First-year Depreciation
50 years	$ 200,000	$ 6,000
30 years	200,000	10,000
20 years	500,000	37,500
10 years	300,000	45,000
5 years	300,000	90,000
10 years[a]	1,500,000	300,000
Total cost	$3,000,000	$488,500

[a] Eligible for investment tax credit

Investment Tax Credit

The tax advantages of a new building are not limited to depreciation. Elements of an owner's building may be eligible for investment tax credits. From time to time government attempts to stimulate economic growth by granting an incentive to firms that construct or purchase capital assets. The presence of this incentive and its amount have varied over time. When such a credit is in existence or expected to come into existence, it should be recognized in the capital budgeting analysis as well as in the project design.

In its present form, the investment tax credit is 10 percent of qualified investment. Qualified investment basically means that the asset is to be used in a production process. The investment tax credit differs from a depreciation expense in that the credit is subtracted from the owner's tax liability rather than used to determine the tax liability. Whereas a dollar depreciation may result in something less than a 50 percent decrease of the tax liability, a dollar investment tax credit results in a dollar decrease in tax liability. In addition, the use of the credit has no effect on the subsequent depreciation deduction.

As the investment tax credit now stands, the owner is entitled to a tax credit of 10 percent of the cost of the eligible asset constructed or purchased in a given year. The expected life of the eligible asset has to exceed 7 years in order to obtain 100 percent usage of the tax credit. The percentage decreases as the life of the eligible asset decreases. For example, for an asset costing $50,000 to construct and having a 10-year life, the owner is entitled to a full 10 percent tax credit or $5000. On the other hand, should the expected life be only 3 years, the owner is entitled to one-third of this or $1667.

Recapture rules exist to handle the case where the asset has been estimated to have a life of 7 or more years only to be replaced after less than 7 years.

Normally a building is not eligible for investment tax credits because it is not directly part of the production process. However, analysis of the elements or components of a building may result in the identification of several types of production elements or components. As an example of a possible investment tax credit element, consider a manufacturing plant with several pieces of production equipment located on the first floor. The slab that holds the equipment might be considered as part of the building and therefore not eligible for investment tax credit. However, it may be that without the need to hold up the production equipment, the slab would only have to be 4 inches thick. However because of the existence of the equipment, 10 inches of concrete may be required to maintain the structural integrity of the building. In effect, the majority of the slab relates to the production equipment. It therefore can be argued that part, if not all, of the concrete slab is eligible for investment tax credit benefits.

Other examples of building elements or components that may be eligible for investment tax credit include plumbing and mechanical work supporting a butcher shop within a grocery store–shopping center, elevators in a building used to transport production material, and even sidewalks and driveways used in transporting production material to and from the manufacturing process.

If a building element can be traced to a production process it may be eligible for investment tax credit. The ability to identify these elements depends on an individual's creativity, analysis skills, and construction and production knowledge. Given these attributes, the construction manager can assist a potential project owner in optimizing investment tax credit benefits and thus lower the cost of the proposed project.

Investment tax credit considerations should also serve as input to design decisions. For example, the fact that movable partition walls are eligible for investment tax credit, whereas fixed walls are ineligible, can affect the design. This is not to say that the designer should design all movable partition walls simply because they may receive better tax treatment. However, the knowledge that movable partitions may qualify should be part of the design considerations. It is obvious that the knowledge of tax implications should be part of the design process.

Capital Gain Treatment

Capital gain treatment relates to the reduced tax rate that applies to the gain received when selling a capital asset. In particular, if the taxpayer sells a business-related building, the building may qualify for favorable capital gain treatment. This type of gain is taxed at an effective rate equal to 40 percent times the rate charged on other income the project owner may have. The only gain that has to be recognized as an *ordinary gain* (that is, a gain that is taxed at the normal rates) is the gain from accelerated depreciation. Excess depreciation is the excess depreciation a project owner has obtained by the use of accelerated methods. The excess depreciation is merely the total depreciation minus the depreciation that would have resulted using the straight-line method. Other than this excess, the owner's gain is given a favorable capital gain treatment. If the feasibility estimate is to be a total analysis of a project, possible capital gain treatment should be recognized.

The gain a project owner would have from the sale of a building would be the difference between the sales price and the building's tax base. Essentially, the tax base is the initial construction cost or purchase cost minus the depreciation taken on the building. Because of the relatively high resale value of buildings, the taxable gain reported may have a significant impact on the economic analysis or feasibility estimate of a project.

Tax Impact of Operating Expenses

Operating expenses and finance-related interest costs are both expenses that can be used to lessen the project owner's income. Finance expenses can be deducted from taxable income in the year that the expenses are incurred. Therefore, although the interest charge for a construction project may be $1,000 a year, in effect the actual interest charge is less when one recognizes the benefits of deducting the expense from total taxable income. For the potential project owner in the 40 percent tax bracket the $1,000 of interest will result in a $400 cash inflow. The result would be an effective interest charge of $600.

Operating expenses can also be deducted in the year in which they occur. Expenses such as utilities, heating costs, everyday repair costs, upkeep costs, and any other related operating costs can be subtracted. Again, the tax value of operating-related costs increases as the taxpayer's income tax bracket increases. For the taxpayer in the 40 percent bracket, this means that every dollar of operating expense is reduced to $0.60. Failure to recognize this benefit can distort the project feasibility estimate.

TIME VALUE OF MONEY CONCEPTS

Earlier reference was made to the relatively long time period that an owner is committed to a construction project. For a single owner, the time may be in excess of 20 years. During this period, both cash inflows and cash outflows affect the economic worth of the project. Time value of money concepts make an early cash outflow more expensive than a later cash outflow, and an early cash inflow worth more than a later cash inflow.

This variation in value necessitates the recognition of the time value of money in the feasibility estimate. Owing to the existence of an interest rate

Time Value of Money Concepts

associated with investing or borrowing money, the value of money to an individual is dependent upon time, that is, the value of a proposed future sum of money must be discounted to determine its present value. When building a project with borrowed money, both the interest rate paid for the money and the time associated with paying it back must be considered to determine the project's true cost.

In this section, several formulas will be given for determining the value of money at different points in time, given a fixed interest rate. These formulas are often referred to as interest formulas. In presenting these interest formulas the following symbols will be used.

   $i.$ interest rate per defined time period
   $n.$ number of time periods
   $P.$ present sum of money
   $S.$ future sum of money
   $R.$ one of a uniform series of end-of-period payments

When the interest paid by the borrower to a lender is proportional to the length of time the money is borrowed, it is referred to as a *simple interest*. The interest paid per period is equal to the amount of money borrowed multiplied by the interest rate per time period. The total amount of interest paid by the borrower to the lender is equal to the amount paid per period multiplied by the number of time periods for which the money is borrowed. Having borrowed an amount of money P from a lender, the borrower pays back the lender a total sum of money S, calculated as follows

$$S = P + Pin = P(1 + in)$$

For example, if a person borrows $1,000 from a lender for a 3-year period at an interest rate of 6 percent, he or she then pays $60 interest at the end of each of the 3 years, in addition to the $1,000 at the end of the last year. These payments are shown on a time scale in Figure 5.5. In the figure, a negative payment refers to money the borrower pays the lender.

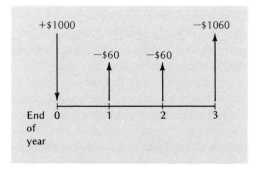

Figure 5:5 Time scale for periodic simple-interest payments

Very few loans are drawn up using simple interest. The more common type of loan uses compound interest. Suppose you lend the bank $1,000. In this case, the bank is the borrower and you are the lender. The loan is such that the bank agrees to pay 6 percent interest compounded yearly for the use of your money. At the end of the first year they credit to your account $60.00 (6% of $1,000). Assuming you withdraw no money from your account, at the end of the second year the bank credits your account $63.60. (6% of $1,060). At the end of the third year they credit your account $67.42 (6% of $1,123.60). The bank is paying you interest on the balance of your account every time period. If after n periods you decide to make a withdrawal of your initial loan and all accumulated interest, you would receive a sum of money given by the formula

$$S = P(1 + i)^n$$

For example, if you withdrew your money after a 5-year period you would receive $1,338 ($1,000$(1 + 0.06)^5$). The payments are shown in Figure 5.6. The positive payment represents the money you receive, the negative payment the money you gave the bank.

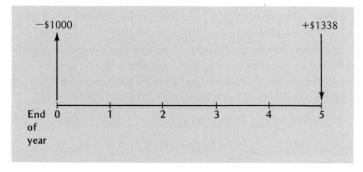

Figure 5:6 Time scale for single-payment of compound amount

It should be obvious that there is a great difference between simple and compound interest. For this reason, it is very important for the borrower and the lender to understand fully the type of interest, the relevant time periods, and the interest rate per time period written into a particular loan. Many people, including owners, have taken out a loan expecting to make certain interest payments, only to discover later they have committed themselves to a larger sum of money. Historically, lenders have been misleading in stating the "real" interest rate associated with a loan to a borrower. For example, a lender may state an interest rate of 1 percent monthly, leading some individuals to believe that on a $1,000 loan, they will only have to pay an interest charge of $10 a year. Of course, this is not correct. A 1 percent monthly interest rate is not equal to a 1 percent annual interest rate. The

Time Value of Money Concepts

government has recently taken steps to alleviate this problem by requiring a lender to state the "true" annual interest rate charged on a loan, rather than a weekly, monthly, or quarterly interest rate.

Regarding the time value of money, an individual may be interested in knowing the worth of a future sum of money at the present time. For example, let us imagine an individual is promised $1,000 five years from the present. The individual realizes that if he or she presently had the $1,000, it could be invested at the present interest rate and accumulate a sum greater than $1,000 by the end of 5 years (at which time he or she would receive the promised $1,000). As a result of this opportunity to invest the money at some positive interest rate today, the future sum of $1,000 is worth less than $1,000 at present.

The present value of a future monetary sum is often referred to as the *present worth* of the money. To determine a future sum's present worth, we need to know how far into the future the sum is to be paid or received, and also the interest rate at which the money could be invested, or in the case of borrowing money, the interest rate associated with borrowing the money. The present worth P of a future sum of money S may be found by using the following formula.

$$P = S(1 + i)^{-n}$$

The term $(1 + i)^{-n}$ is known as the single-payment present worth formula or factor. It assumes a constant i over the relevant time period n. Observe that the single payment present worth factor is merely the inverse of the single payment compound interest factor, that is, finding the future worth of a present sum of money is the reverse of finding the present value of a future sum.

Returning to the problem of finding the present worth of $1,000 to be received 5 years from now, let us assume the individual could invest the money at an interest rate of 6 percent. The present worth of the money is equal to $747.30 ($1,000 $(1 + 0.06)^{-5}$). Figure 5.7 shows present worth on a time scale.

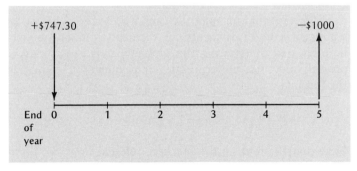

Figure 5:7 Time scale of single-payment by present worth formula

By using the single payment compound amount factor and the single payment present worth factor, the value of a sum of money may be found at any desired point in time. However, to facilitate the actual calculation, which may involve shifting money on a time scale when there are several payments, several other interest formulas or factors have been derived. One of these factors is the *capital recovery interest factor*. This is useful in determining the equal payments required in a loan in which the borrower is to only pay interest on the unpaid balance. For example, suppose an individual is to repay a $1,000 loan in 5 equal end-of-the-year payments, paying a 6 percent interest rate on the unpaid balance. The equal payments R he or she is to make may be obtained from the following formula.

$$R = P \left[\frac{i(1+i)^n}{(1+i)^n - 1} \right]$$

The term in brackets is known as the capital recovery factor. Note that the factor is merely the result of summing several single-payment compound amount factors. In the above example, the interest factor is found to be 0.2374. The equal payments R to be made are found to equal $1,000(0.2374) or $237.40. The payments involved are shown on a time scale in Figure 5.8.

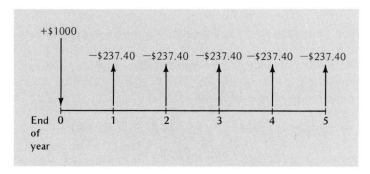

Figure 5:8 Time scale of payments by capital recovery factor

Observe that in using the capital recovery factor we are actually finding a uniform series of payments equivalent to a present monetary sum. We may be interested in reversing the problem; that is, we may want to know the present value of a uniform series of payments. This may be found by using the formula

$$P = R \left[\frac{(1+i)^n - 1}{i(1+i)^n} \right]$$

Note that the term in the brackets, referred to as the *uniform series present worth factor*, is merely the inverse of the capital recovery factor.

Time Value of Money Concepts

The uniform series present worth factor is useful in determining the present worth of a uniform series of payments. For example, let us suppose an individual is promised $1,000 at the end of every year for 5 years. Since the available money can be invested at a 6 percent interest rate, this promised sum of $5,000 (5 × $1,000) is worth less than $5,000 at the present time. The uniform series present worth factor for $i = 0.06$, $n = 5$, is found to be 4.2124. Thus, the present worth of the payments is equal to $4,212.40 ($1,000 × 4.2124). These payments are shown on a time scale in Figure 5.9.

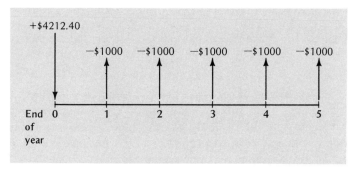

Figure 5:9 Time scale of payments by uniform series present worth factor

Two other interest formulas or factors are of interest here. These interest factors define the relationships that exist between a future sum of money S and a set of uniform series of payments R. These relationships may be derived from the single payment compound amount interest factors. One of these interest factors, the *uniform series compound amount factor*, determines what equal-value, end-of-period payments, invested at an interest rate i and earning interest immediately upon investment, will sum to after n payment periods. The amount accumulated, which consists of the sum of the uniform payments and the earned interest, may be determined by

$$S = R \left[\frac{(1+i)^n - 1}{i} \right]$$

The term in the brackets is referred to as the uniform series compound amount factor. Suppose an individual invests $1,000 at the end of every year into a savings account that pays 6 percent compounded interest. The money starts drawing interest as soon as it is invested. To determine the amount of money accumulated after 5 years, we calculate the uniform series compound amount factor for $i = 0.06$, $n = 5$, and obtain a value of 5.637. Thus, the money accumulated is the product of $1,000 times 5.637, or $5,637. This is shown on the time scale in Figure 5.10.

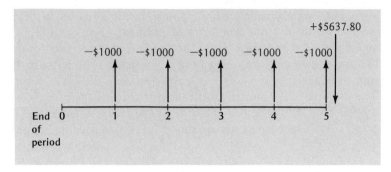

Figure 5:10 Time scale of payments by uniform series compound amount

The inverse of the uniform series compound amount factor is the *sinking fund factor*. The sinking fund factor determines which uniform series of payments must be made at a given interest rate to accumulate a stipulated future sum S after a given number of payment periods. For example, assume an individual desires to invest a uniform series of payments (each to be made at the end of the year at an interest rate of 6 percent), so that there is $1,000 in savings at the end of 5 years. This uniform series of payments may be derived from the formula

$$R = S\left[\frac{i}{(1+i)^n - 1}\right]$$

The term in the brackets is referred to as the sinking fund factor. For the above example, the sinking fund factor equals 0.1774. Thus, the uniform end-of-the-year payments that the individual must invest in order to accumulate $1,000 after 5 years are equal to $1,000 multiplied by 0.1774, or $177.40. These payments are shown on a time scale in Figure 5.11.

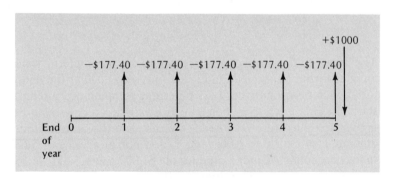

Figure 5:11 Time scale of payments by sinking fund factor

Discounted Cash Flow Analysis

It should be observed that the value of any interest factor depends only on the relevant interest rate i and the number of payments n. From this point on, the entire formula will not be written out for a given interest factor. Rather, a shortened notation will be substituted. Single payment compound amount will be abbreviated as (SPCA, $i = x$, $n = y$), single payment present worth as (SPPW, $i = x$, $n = y$), uniform series present worth as (USPW, $i = x$, $n = y$), capital recovery as (CR, $i = x$, $n = y$), sinking fund as (SF, $i = x$, $n = y$), and uniform series compound amount as (USCA, $i = x$, $n = y$). The values of x and y correspond to the given interest rate and the number of payment periods.

The values of the various interest factors as a function of i and n are available in interest tables. The use of such interest tables eliminates the need to solve the interest formula. A set of interest tables is presented in appendix C.

DISCOUNTED CASH FLOW ANALYSIS

The most general technique available to a construction manager and owner for evaluating project feasibility is the *discounted cash flow method*. This method is alternatively referred to as the *rate of return method*. Using this method, a capital expenditure is viewed as the acquisition of a series of future net cash flows consisting of two elements: the return of the original outlay and the net income or cost savings from the capital expenditure. Should the present value of the net cash inflows discounted at a desired rate of return be in excess of the initial capital expenditure required, then the investment may be warranted. Alternately, two proposals can be compared on the basis of their respective differences in discounted net cash inflows versus required initial investment.

Let us assume a potential project owner is considering a project with an initial cost of $60,000. The estimated annual cash flow from rent is $9,000, and the increased cash flow from depreciation is $5,000. Thus the total annual cash inflow is $14,000. To determine the feasibility of the capital expenditure using the discounted cash flow method, the future incomes or cash flows are discounted back to the present time and compared to the initial capital expenditure of $60,000. The future cash flows are discounted by calculating their present worths at a specified interest rate. The interest rate to be used in the calculation is the one for which the firm can invest the initial capital expenditure if it does not build the project, or at which the firm borrows the money to build the building. Let us assume that the latter is true in the example, and the effective interest rate, including closing costs, and so on, is 15 percent. The net cash inflow discounted at 15 percent is calculated in Figure 5.12. The discounted sum of the inflows exceeds the initial cost of $60,000. This means that there is a positive economic return from the building.

Year	Net Cash Flow	Present Value of $1 at 15%	Present Value of Net Cash Flow
1	$14,000	0.869	$12,166
2	$14,000	0.756	$10,584
3	$14,000	0.657	$ 9,198
4	$14,000	0.572	$ 8,008
5	$14,000	0.497	$ 6,958
6	$14,000	0.432	$ 6,048
7	$14,000	0.376	$ 5,264
8	$14,000	0.327	$ 4,578
9	$14,000	0.284	$ 3,976
10	$14,000	0.247	$ 3,458
11	$14,000	0.215	$ 3,010
12	$14,000	0.187	$ 2,618
			$75,866

Figure 5:12 Discounted cash flows at 15 percent

The use of the discounted cash flow or rate of return method is not limited to determining if the discounted net cash inflows exceed the initial expenditure for an investment alternative. The method can be used to determine the actual rate of return associated with an investment. For example, let us assume that the owner contemplating the $60,000 project would like to know the true rate of return associated with it. Because the total net cash inflows sum to $168,000, whereas the initial expenditure is only $60,000, it follows that the rate of return is greater than zero. It can also be concluded from the calculations in Figure 5.12 that the rate of return on the initial expenditure exceeds 15 percent. This follows from the fact that the sum of the net cash inflows, discounted at an assumed interest rate of 15 percent, exceeds the initial expenditure of $60,000. In effect, this means that had the owner invested the $60,000 at a 15 percent compounded annual interest rate, it would have received fewer cash inflows than those received from investment. Thus the return associated with the $60,000 project must be greater than 15 percent.

Having decided that the actual rate of return is in excess of 15 percent, the next step in determining the actual rate of return involves trial-and-error calculations. The complex mathematical nature of interest formulas results in the necessity of using trial and error. A rate of return is selected on the basis of eliminating unfeasible values (in this case all values less than or equal to 15 percent are unfeasible), and from analyzing previously calculated discounted net cash inflows versus initial expenditures. For example, from Figure 5.12 it was determined that the sum of the discounted net cash inflows equals $75,866. This sum, discounted at 15 percent, is quite a bit in excess of the $60,000 investment. On this basis it follows that the next trial-

Discounted Cash Flow Analysis

and-error iteration should be quite a bit in excess of 15 percent. Perhaps 25 percent would be a good estimate. The sum of the discounted net cash inflows at 25 percent is calculated in Figure 5.13.

Year	Net Cash Flow	Present Value of $1 at 25%	Present Value of Net Cash Flow
1	$14,000	0.800	$11,200
2	$14,000	0.640	$ 8,960
3	$14,000	0.510	$ 7,168
4	$14,000	0.410	$ 5,734
5	$14,000	0.328	$ 4,588
6	$14,000	0.262	$ 3,669
7	$14,000	0.210	$ 2,936
8	$14,000	0.168	$ 2,349
9	$14,000	0.134	$ 1,879
10	$14,000	0.107	$ 1,504
11	$14,000	0.086	$ 1,203
12	$14,000	0.069	$ 962
			$52,152

Figure 5:13   Discounted cash flows at 25 percent

From the calculation in Figure 5.13 it is determined that the sum of the net cash inflows discounted at 25 percent is less than $60,000. This means that had the firm invested $52,152 in the project and received the annual cash inflows of $14,000 for 12 years, the effective rate of return on the investment would be 25 percent. However, because the firm had to invest more than $52,152 (the project required an expenditure of $60,000), its actual rate of return is less than the 25 percent.

We have now bounded the solution. The actual rate of return is between 15 and 25 percent. If it is desirable to determine the actual rate of return more accurately, the next iteration would be carried out at a rate somewhere between 15 and 25 percent. The next trial might be made at 20 percent. This is done in Figure 5.14.

Although the sum of the net cash inflows discounted at 20 percent does not exactly equal $60,000 it is relatively close. Any further calculation of the actual rate of return may not be beneficial in that there is a degree of uncertainty associated with the expected annual net cash inflows. More likely than not the rate is close to 21 percent.

The trial-and-error method has resulted in the determination of approximately a 21 percent rate of return for the project. Having calculated this, it can be compared to the expected rate of return for alternative investments to determine if the project is warranted. Note that although the actual rate of

Year	Net Cash Flow	Present Value of $1 at 20%	Present Value of Net Cash Flow
1	$14,000	0.833	$11,662
2	$14,000	0.694	$ 9,716
3	$14,000	0.579	$ 8,106
4	$14,000	0.482	$ 6,748
5	$14,000	0.402	$ 5,628
6	$14,000	0.335	$ 4,690
7	$14,000	0.279	$ 3,906
8	$14,000	0.233	$ 3,262
9	$14,000	0.194	$ 2,716
10	$14,000	0.162	$ 1,890
11	$14,000	0.135	$ 2,268
12	$14,000	0.112	$ 1,568
			$62,160

Figure 5:14 Discounted cash flows at 20 percent

return was determined using a trial-and-error method, it could have been determined in a more straightforward manner. Because each of the expected annual cash inflows was equal in amount, the rate of return could have been solved from the following equation.

$$\$60,000 = (\$14,000)(\text{Uniform series present worth factor})$$

Using interest formula tables, the value of i can be found to solve the equation. This interest rate is the rate of return. This method can only be used when the expected inflows are equal. More often than not they vary, because expected revenues and expenses vary, and when this is the case the trial-and-error method must be used.

OFFICE BUILDING EXAMPLE (NO COMPONENTIZING)

The purpose of this section is to give an example of a feasibility estimate. The relevant factors and terminology are a function of the type of project being considered. As such, it is impossible to present an all-encompassing example. However, the procedures and objectives of the feasibility study are essentially the same, independent of type or size of project. Thus the techniques illustrated here are applicable to many types of projects.

It will be assumed that a firm is considering constructing an office building with the intention of leasing space. Based on its available budget, lot size, and familiarity with similar projects, the potential investor has decided on a three-story building with approximately 12,500 square feet. At the

Office Building Example (No Componentizing)

time of the feasibility estimate, little has been determined about building quality other than the fact that a brick building is preferred. The project is to be conventionally financed and is to be built in a metropolitan area.

The initial cost of the project can be approximated by using costs per square foot data published in cost books. A 12,500 square foot building will cost $80.00 per square foot or $1 million according to our cost book. This cost is for the *improvements* on the land. To it must be added the land cost. This cost is usually more certain than the improvement costs because a predetermined price can be set by means of an *option* contract with the landowner. Such a contract gives the potential investor a right to purchase the land in question for a set amount of money within a given time period—say 2 months. For this right the investor gives the landowner an amount of money that will be forfeited if the investor fails to exercise the purchase right. Should the investor purchase the land, the option money usually is credited to the purchase price. For our example, let us assume the investor has a right to purchase the required land for $100,000. Thus, the initial office building costs are approximated at

Land	$ 100,000
Improvements	1,000,000
Initial Costs	$1,100,000

Let us assume that the cost of money is such that it will be possible for the investor to secure an 80 percent conventional financing loan at 15 percent interest per year payable over 20 years. The other 20 percent of the initial costs are to be financed by equity capital. Thus the investor borrows 80 percent of the $1.1 million or $880,000.

By using the amortization tables in appendix C (the tables can be developed using time value of money formulas discussed in the previous section), it is possible to calculate the equal monthly or yearly payment that retires the loan and reflects the interest charged by the lender. In a feasibility estimate, this monthly or yearly payment is commonly referred to as the *debt service*. From appendix C, $140,588.80 is determined as being the annual payment required by the company to repay the $880,000 at 15 percent.

The operating expense component of the project is commonly expressed as a percentage of the construction costs of the building. Operating expenses normally range from 6 to 30 percent of construction cost, depending on the type of building. Let us assume that it is expected that the initial operating expense (maintenance and repair) for the office building is 10 percent of the construction costs. Based on anticipated construction costs of $1 million, the initial operating expenses are estimated to be $100,000.

A market analysis must be made of the immediate project vicinity to establish an estimate of the rental market. A primary part of the market analysis consists of a comparison of rental or leasing rates of comparable types of buildings in the surrounding area. For our example, let us assume that the market analysis yields an anticipated leasing rate of $2.50 per square foot per month.

Although the gross square footage of the proposed office building is 12,500, interior walls, doorways, columns, and so on will reduce the leasable space. The rule of thumb in office buildings is that a building is efficiently designed if the leasable area approaches 80 percent of the gross area. Assuming 80 percent leasable space, a 7 percent vacancy allowance, and an expected leasing rate of $2.50 per square foot per month, the gross possible income is calculated as follows.

$$\begin{aligned}
\text{Gross area} &= 12{,}500 \text{ sq. ft.} \\
\times\ .80 \text{ leasable space} &= 10{,}000 \text{ sq. ft.} \\
-\ .07 \text{ vacancy allowance} &= 9{,}300 \text{ sq. ft.} \\
\times\ \$2.50 \text{ per sq. ft.} &= \$23{,}250 \text{ per month} \\
\times\ 12 \text{ months} &= \$279{,}000
\end{aligned}$$

As noted earlier, operating expenses will be assumed as $100,000 for the first year. Naturally, the operating expenses will increase from one year to the next. However, they will be compensated for by equal increases in leasing rates.

The most accurate means of recognizing all relevant factors affecting the rate of return of a project is through application of the *discounted rate of return method*. The fundamentals of the discounted rate of return method were discussed in an earlier section.

Although the discounted cash flow method gives an accurate rate of return calculation, this accuracy can only be obtained if there is sufficient information. In addition, to carry out a thorough analysis, assumptions must be made. For example, for the office building project being analyzed, the owner's decision to sell the project after a number of years, the investor's tax rate, and the depreciation method to be used must be input.

Let us assume the office building investor is in the 40 percent tax bracket, that the investor is planning to sell the project after 7 years, and that the investor would like to use a depreciation method that maximizes the project's rate of return. The fact that the owner plans to sell the project before it has no value requires an estimate of the selling price. This estimate is needed to reflect properly the payment to the owner. The estimated future value of a building is often determined by assuming a constant appreciation (or depreciation) rate for the building. This rate is unrelated to the depreciation rate used to reduce the value of the building for tax purposes. For the office building in this example we will assume an annual appreciation rate of 1 percent.

Having made the assumptions, we will proceed to determine the annual cash inflow or outflow for each year. As indicated earlier, the gross possible income (the rental income) and the operating expense will vary from year to year. However, since the investor plans to increase rental rates to reflect increased operating expenses, it is the net of the rental income and operating expenses that is of concern, and this amount can be assumed constant.

Initially, the investor in the office building has an outflow of $220,000. This represents the invested equity in the project. The annual cash inflows from owning and operating the office building for the first year are illustrated in Figure 5.15. The cash-flow calculation is initiated by first listing the gross possible income (the lease income). The $279,000 shown is minus the vacancy allowance.

Pro Forma
Annual Cash Results for Office Building

A.	Gross Possible Income	$279,000.00
B.	Operating Expenses	−100,000.00
C.	Net Operating Cash Income	$179,000.00
D.	Less Mortgage of $880,000 times Debt Service Constant of 0.15976	−140,588.80
E.	Net Cash Flow	$ 38,411.20
F.	Depreciation, Interest, and Tax Effects	

1. Gross Possible Income	$279,000.00	
2. Operating expenses	100,000.00	
3. Depreciation-Construction costs of $1 million at 150% straight-line depreciation with life of 40 years	−37,500.00	
4. Interest cost (part of the debt service)	−132,000.00	
Net Operating Gain	$ 9,500.00	
Tax Liability (40% bracket) ($9,500.00 × 0.40)		− 3,800.00
G. Total Cash Inflow After Taxes		$ 34,611.20

Figure 5:15   First year cash inflow

The operating expenses include annual utilities, maintenance and repairs, insurance, real estate, taxes, advertising, legal and accounting costs, and caretaker/manager fees. Assuming operating expenses of 10 percent of initial construction costs of $1 million, $100,000 is shown in Figure 5.15 as the first year's operating expense. The difference between the gross possible income and the operating expense is shown as the *net operating cash income*. It is the cash inflow to the investor before consideration of the debt service, depreciation, and tax effects.

The debt service amount of $140,588.80 is subtracted from the net operating cash income. The constant of 0.15976 used to determine the debt service is a factor that equates a 15 percent interest rate with a payback period of 20 years. Subtracting the $140,588.80 from the $179,000 leaves a net cash flow of $38,411.20. The calculated net cash flow does not reflect depreciation, or tax effects.

Part F of Figure 5.15 reflects the effect of depreciation, interest, and taxes. The calculation starts with restating of the gross possible income. In order to calculate the actual operating gain or loss for tax purposes, operating expenses, depreciation, and interest must be subtracted from the gross possible income. The operating expense was previously calculated as $100,000. This is shown in line 2.

Depreciation is an allowable tax expense. However, depreciation is not a cash expense to the firm. The result is that depreciation provides the potential investor a noncash expense that reduces taxes. Since the office building investor wants to use the depreciation method that maximizes the project's rate of return, the maximum allowable rate for office buildings under current law of 150 percent of straight-line depreciation is taken. This is sometimes referred to as 150 percent declining-balance depreciation. Current tax laws also dictate the minimum number of years over which an asset can be depreciated. Depending upon the type of building, the minimum number of years ranges from 20 to 50. It is typical to depreciate office buildings of the type in this example over 40 years. Given these assumptions, the calculated depreciation for the first year of the office building is $37,500.

To pay back the mortgage on the office building, we previously calculated an annual debt service of $140,588.80. This debt service was made up of a payback of principal and interest. Only the interest is deductible for calculating taxable income. As such, the interest cost must be broken out of the debt service to calculate the effect of interest on cash flow. In early years the interest expense is large, while the principal payback is small. In particular, the interest cost for the first year is 15 percent times the mortgage. This yields an interest expense of $132,000 for the first year.

Subtracting the operating expenses, depreciation, and interest expense from the gross possible income gives a net operating tax gain of $9,500. It should be noted that in the case of some project investments, the effect of the depreciation and interest expense may result in a net operating loss. This net operating loss is desirable in that it results in a reduction of taxes to the point that the tax effect actually adds to the net cash flow for the investor. In our example, the gross possible income is so large that even the effects of depreciation and interest do not reduce the net operating income to a negative number. However, these expenses still reduce the tax liability.

Office Building Example (No Componentizing)

Given that the investor is in the 40 percent income tax bracket, the tax liability is calculated by multiplying 40 percent times the net operating gain of $9,500. This results in a tax liability of $3,800. This liability must be subtracted from the net cash flow in order to calculate the total cash inflow after taxes. As is shown in Figure 5.15, this results in a total cash inflow after taxes of $34,611.20. This is the total cash inflow at the end of the first year for the office building. This inflow will change as the depreciation and interest expenses change. This will change the tax liability, which will have the effect of decreasing total cash inflow.

To calculate the total cash inflow for years 2 through 7, we must be able to determine the actual depreciation charge for each year and the actual interest expense. These calculations are made in Figures 5.16 and 5.17. In Figure 5.16, the depreciation charge for any one year is determined by multiplying the book value at the end of the previous year by 0.0375. This factor is determined from the fact that the depreciation rate is 150 percent of straight-line and that the depreciation life is 40 years. Mathematically, 0.0375 is obtained by dividing 40 into 1, and multiplying the result by 1.5. The 1.5 reflects the 150 percent straight-line depreciation rate. In Figure 5.16 the book value after deducting the depreciation charge for each year is shown in the last column.

End of Year	Depreciation Charge	Book Value
0		$1,000,000.00
1	$37,500.00	962,500.00
2	36,093.75	926,406.25
3	34,740.23	891,666.02
4	33,437.47	858,228.55
5	32,183.57	826,044.98
6	30,976.69	795,068.29
7	29,815.06	765,253.23

Figure 5:16 Depreciation charge and book value for improvements

The calculation of the interest payment is shown in Figure 5.17. The debt service charge for each year is constant. This payment is $140,588.80, as shown in Figure 5.17. Note that the debt service includes an interest and a principal payment. The interest payment is calculated by multiplying the interest rate of 15 percent by the mortgage at the end of the previous year. The remainder after subtracting the interest payment from the debt service is the principal payment. As can be seen in Figure 5.17, the interest payment decreases as the principal payment increases.

Having calculated the depreciation expense and interest expense for years 1 through 7, we can now calculate the total cash inflow for each of these

End of Year	Debt Service	Interest Payment	Principal Payment	Mortgage
0				$880,000.00
1	$140,588.80	$132,000.00	$ 8,588.80	871,411.20
2	140,588.80	130,711.68	9,877.12	861,534.08
3	140,588.80	129,230.11	11,358.69	850,175.39
4	140,588.80	127,526.30	13,062.50	837,112.89
5	140,588.80	125,566.93	15,021.87	822,091.02
6	140,588.80	123,313.65	17,275.15	804,815.87
7	140,588.80	120,722.33	19,866.47	784,949.40

Figure 5:17 Debt service breakdown

years. Previously we determined the total cash inflow at the end of year 1. Note that the depreciation expense and the interest expense have the effect of changing the tax liability for each of the years. The operating expense and the gross possible income or rental income will be assumed constant. The net operating gain for each of the years shown in Figure 5.18 was calculated by subtracting the operating expense and depreciation expense from the gross possible income for each year. The tax shown in Figure 5.18 was determined by multiplying 40 percent by the net operating gain for that year. Finally, the total cash inflow is calculated by subtracting the tax liability from the net cash flow of $38,411.20 previously calculated.

End of Year	Operating expense	Depreciation	Interest	Net Operating Gain	Tax	Total Cash Inflow
1	$100,000.00	$37,500.00	$132,000.00	$ 9,500.00	$ 3,800.00	$34,611.20
2	100,000.00	36,093.75	130,711.68	12,194.57	4,877.83	33,533.37
3	100,000.00	34,740.23	129,230.11	15,029.66	6,011.86	32,399.34
4	100,000.00	33,437.47	127,526.30	18,036.23	7,214.49	31,196.71
5	100,000.00	32,183.57	125,566.93	21,249.50	8,499.80	29,911.40
6	100,000.00	30,976.69	123,313.65	24,709.66	9,883.86	28,527.34
7	100,000.00	29,815.06	120,722.38	28,462.56	11,385.02	27,026.18

Figure 5:18 Cash inflows for years 1–7

An additional net cash inflow occurs at the end of the seventh year from the sale of the office building. The calculation of the net cash inflow from the sale is shown in Figure 5.19. The calculation starts with listing the sale price. The anticipated sale price is determined from our previous assumption of a 1 percent appreciation rate per year for the land and the improve-

Office Building Example (No Componentizing)

```
A. Cash for Sale
   1. Sales price
      (1% appreciation)              $1,179,348.80
   2. Less 5% commission                 58,967.40
   3. Gross cash                     $1,120,381.40
   4. Less morgage balance              784,949.40
   5. Adjusted cash income          $   335,432.00
B. Taxes to be Paid
   1. Gross cash                     $1,120,381.40
   2. Less book value                   865,253.23
   3. Taxable gain                  $   255,128.17
   4. Taxable as ordinary income         59,746.80
   5. Taxable as capital gain           195,384.37
   6. Total tax due                      55,160.20
C. Net Cash from Sale
   1. Adjusted cash income          $   335,432.00
   2. Less taxes due                     55,160.20
   3. Net cash                      $   280,271.80
```

Figure 5:19 Cash flow from sale

ments. This assumption results in an anticipated sale price of $1,179,348.80. We assume that the owner of the office building will incur a 5 percent commission cost to sell the building. This cost is subtracted from the anticipated sale price to yield gross cash of $1,120,381.40. The mortgage balance at the end of the seventh year for the loan is calculated in Figure 5.17 as $784,949.40. When the office building is sold, this mortgage balance will have to be paid to the lending institution. After paying this mortgage balance the adjusted cash flow to the investor is $335,432.

Taxes on the gain from the sale of the office building will reduce the adjusted cash income. The calculation of the tax liability is shown in Figure 5.19. The calculation is made by subtracting the book value at the end of the seventh year from the gross cash received from the sale. The book value at the end of the seventh year was calculated in Figure 5.16. The result is a taxable gain of $255,128.20. Current tax laws break this tax into two liability categories, an ordinary gain and a capital gain. A capital gain has the advantage of being taxed at 40 percent of the rate that an ordinary gain would be taxed. Current tax laws are such that the only gain that must be recognized as an ordinary gain is the excess depreciation to the date of the sale. Excess depreciation is defined as depreciation deductions above what would have been determined using the straight-line method. For the example in question, the ordinary gain is calculated as $59,746.80. The capital gain was calculated as $195,384.37.

The actual tax liability is calculated using the tax rate that was previously assumed for the office building investor of 40 percent. As such, the total tax liability is determined by multiplying 40 percent times the ordinary gain of $59,746.80 and adding this amount to the product of 16 percent times the capital gain of $195,384.37. This yields a total tax due of $55,160.20. Finally, the net cash flow to the investor from the sale of the office building at the end of the seventh year is determined by subtracting the tax liability from the adjusted cash income previously calculated. This calculation is shown in Figure 5.19.

We have now calculated the cash outflows and inflows for the period for which the investor is affected by the project. These cash outflows and inflows are best illustrated on a time scale as in Figure 5.20. The arrows shown on the time scale indicate a cash outflow or a cash inflow; arrows pointing upward indicate a cash outflow and the arrows pointing downward represent net cash inflow. For this example there is only one net cash outflow. This occurs at the time of the initial cash investment in the project. Note that after the seventh year there are two arrows shown indicating two net cash inflows. They represent the cash inflow from operating the office complex in the seventh year and the cash inflow from the sale of the office building after the seventh year.

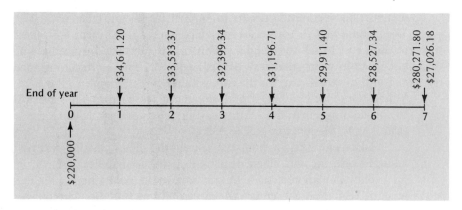

Figure 5:20 Time scale of cash inflows and outflows for non-componentized $220,000 investment

At this point the analysis has yielded a net cash inflow or outflow for each year. This provides the potential investor a tool for evaluating the feasibility of the project. The rate of return for any one year can be determined by dividing the cash inflow for that year by the equity investment. For exam-

Office Building Example (No Componentizing)

ple, for the third year a net cash inflow was determined as $32,399.34. Thus the rate of return is $32,399.34 divided by $220,000. This calculation yields a rate of return for the third year of approximately 14.7 percent.

The error in calculating the rate of return in this manner is that such a calculation neglects the time value of money. Future net inflows of money are not worth as much as if the cash inflows were received at the present time. A true rate of return calculation should discount future net cash inflows to reflect the time they are received. To discount future cash inflows or outflows in a rate of return calculation, the analysis must include the interest formulas discussed in a previous section. The analysis consists of determining the interest rate at which future sums of money are discounted back to the present time to equal the initial investment. Let us look at Figure 5.20. We see there are cash inflows at the end of each of the years 1 through 7. If we can determine the interest rate i at which the discounted value of the sum of each inflow equals the initial investment of $220,000, then we will have found the true rate of return for this investment.

This is essentially a trial-and-error analysis. First we guess at the rate of return. The initial guess can be guided by looking at the magnitude of the future cash inflows versus the initial investment. If one were to sum all of the future cash inflows they would equal $497,477.34. This is considerably larger than the initial investment of $220,000.00. It follows that because the cash inflows are considerably in excess of the initial investment, that a fairly large rate of return seems evident. Let us start the analysis by assuming a rate of 15 percent. The cash inflows in Figure 5.21 are multiplied by interest factors that reflect the year at which the cash inflow is received. A sum of $237,311.35 is determined as the present worth of the future cash inflows. This value exceeds the $220,000 initial investment. A comparison of these two numbers shows the true rate of return for the office building to be greater than 15 percent. This conclusion is reached by reasoning that if the investor contributed $237,311.35 in equity for the office building that he would in fact have re-

Year	Cash Inflows	SPPW $i = 15\%$, $n = x$	Discounted Value
1	$ 34,611.30	0.8696	$ 30,097.99
2	33,533.77	0.7561	25,354.88
3	32,399.34	0.6575	21,302.57
4	31,196.71	0.5718	17,838.28
5	29,911.40	0.4972	14,871.95
6	28,527.34	0.4323	12,332.37
7	307,297.98	0.3759	115,513.31
			$237,311.35

Figure 5:21 Discounted cash inflows at 15 percent

ceived the cash inflow illustrated in Figure 5.20. However, the owner of the office building did not have to invest $237,311.35 in order to receive the cash inflow. He only had to invest $220,000. Thus the rate of return exceeds 15 percent.

We now know that the rate of return is greater than 15 percent. However, the question remains as to how great the true rate of return is. We require another guess. Let us assume an interest rate of 20 percent. By so doing, it is likely that we will bound a solution. That is, should we find an interest rate of 20 percent is too great, at least we will know that the rate of return is between 15 and 20 percent.

Year	Cash Inflows	SPPW $i = 20\%$, $n = x$	Discounted Value
1	$ 34,611.30	0.8333	$ 28,841.60
2	33,533.77	0.6944	23,285.85
3	32,399.34	0.5787	18,749.50
4	31,196.71	0.4823	15,046.17
5	29,911.40	0.4019	12,021.39
6	28,527.34	0.3349	9,553.81
7	307,297.98	0.2791	85,766.87
			$193,265.19

Figure 5:22 Discounted cash inflows at 20 percent

The discounted cash inflows at 20 percent are shown in Figure 5.22. The mathematics is the same as in Figure 5.21. The only difference is that the interest factor is now 20 percent. As in Figure 5.22, the sum of the present worths of the discounted cash inflows is $193,265.19. This is less than the initial investment of $220,000.00. This means that had the owner invested $193,265.19 in equity he would have obtained the cash inflow shown in Figure 5.20. However, the investor in the office building had to invest more than this amount. This means that the true rate of return is less than 20 percent.

We have now bounded a solution. Naturally, the next guess should be between 15 and 20 percent. We will make our next guess at an interest rate of 17 percent. The analysis is shown in Figure 5.23. It can be seen that the present worth of the discounted cash inflows discounted at 17 percent results in a sum of $218,109.32. This is very close to the initial investment of $220,000. Thus the true rate of return is slightly less than 17 percent. However, we will not carry the solution any closer than we have at this point. Recognizing all the uncertainties of the various cash inflows, further analysis is not justified.

The discounted cash flow analysis of the office building project is now

Office Building Example (Using Componentizing)

Year	Cash Inflows	SPPW $i = 17\%$, $n = x$	Discounted Value
1	$ 34,611.30	0.8547	$ 29,582.28
2	33,533.77	0.7305	24,496.42
3	32,399.34	0.6244	20,230.15
4	31,196.71	0.5336	16,646.56
5	29,911.40	0.4561	13,642.59
6	28,527.34	0.3898	11,119.96
7	307,297.98	0.0332	102,391.36
			$218,109.32

Figure 5:23 Discounted cash inflows at 17 percent

complete. The result of the analysis has been the determination of the return the investor will receive from the equity investment in the project. This rate of return can be compared to alternative investments that may be available. Included in these alternative investments are such things as stocks, treasury bills, bonds, and simple savings and loan passbook accounts. Other investments might include other business ventures such as investments in new products, in new lines of business, or perhaps other types of construction projects. The analysis can also be used as a basis for evaluating design modifications. The point to be made is that the rate of return that has been calculated reflects all relevant financial considerations so that it is the true rate of return to the investor.

OFFICE BUILDING EXAMPLE (USING COMPONENTIZING)

The feasibility analysis of the office building illustrated in the previous section assumed that the building was to be depreciated as a single entity with a 40-year depreciable life. An approximate rate of return of 17 percent was calculated. As noted in an earlier section of this chapter, the cash flow received from depreciation can be speeded up through a componentized depreciation analysis of a building. Individual components of a building may have a useful life less than that of the whole building. By depreciating the components separately, the depreciation benefits as well as the rate of return can be improved.

To illustrate the benefits of a componentized analysis of a building, let us reconsider the same office building. Let us assume that the construction manager, using its construction and estimating skills, performs a componentized analysis of the building. This would be performed by analyzing each of the systems of the building, such as the structural, electrical, mechanical, plumbing, and finish systems. The componentization of the electrical system is illustrated in Figure 5.24. The costs shown include a proration of the indirect costs, such as the architect/engineer costs.

25-year life		20-year life	
Switchboard	$ 1,022.42	Transformer	$ 949.90
Power mains	1,409.84	Feed wiring	5,079.17
Switch gear	12,488.05	Distribution panels	2,707.53
Wiring & conduit	33,550.14	Telephone conduit	2,351.45
	$48,470.45	Security lighting	2,454.31
		Motor control centers	37,083.85
		Motor wiring	19,289.76
		Emergency plant	781.10
			$70,697.07
15-year life		10-year life	
Branch wiring	$16,135.85	Installed air connectors	$ 60.03
Lighting fixtures	24,240.30	Electrical connectors	413.65
Air conditioning connectors	200.22	Electric sign connectors	113.28
Circulating pump	36.60		$586.96
Visitors display	624.20		
Reflecting illumination	218.64	Summary	
Case housing	934.90		
Air conditioner exhaust	763.38	25-year life	$ 48,470.45
Tank connections	79.30	20-year life	70,697.07
Paging system	838.89	15-year life	45,233.46
Intercom system	1,161.18	10-year life	586.96
	$45,233.46	Subtotal	$164,987.94
		A/E proration	8,249.40
		Indirect cost proration	19,798.55
		Total	$193,035.89

Figure 5:24   Analysis of electrical expenditures

Although the componentization of other systems is not shown, the summary results of the componentized depreciation are illustrated in Figure 5.25. The summary includes component life groups of 50, 25, 20, 15, and 10 years. The 50-year group recognizes that the building frame, apart from other components, may have a useful life greater than the composite life of the building depreciated as a single entity (that is, 40 years).

Office Building Example (Using Componentizing)

Life (years)	Cost
50	$ 112,424.52
25	245,696.33
20	344,823.24
15	212,211.21
10	84,844.70
Total Construction Cost	$1,000,000.00

Figure 5:25 Summary of componentizing

The feasibility estimate of the office building is performed in Figures 5.26 through 5.33. The only difference in the calculation process relative to the noncomponentized analysis is the annual depreciation amount. The annual depreciation amounts are calculated in Figure 5.27. The annual depreciation for each group is calculated by using a depreciation rate 150 percent of the group's straight-line rate. For example, for the 50-year group, a 3 percent depreciation rate is determined by multiplying the 2 percent straight-line rate by 1.5.

Pro Forma Annual Cash Benefits
Office Building-Componentized

A. Gross Possible Income		$279,000.00
B. Operating Expenses		−100,000.00
C. Net Operating Cash Income		$179,000.00
D. Less Mortgage of $880,000 times Debt Service Constant 0.15976		−140,588.80
		$ 38,411.20
E. Depreciation, Interest, and Tax Effects		
1. Gross possible income	$279,000.00	
2. Operating expense	−100,000.00	
3. Depreciation-construction costs of $1,000,000 at componentized rate (See Figure 5.23)	− 77,924.06	
4. Interest cost (part of the debt service)	−132,000.00	
Net Operating Gain	−$30,924.06	
Tax Liability (40% bracket) (−$30,926.04 × 0.40)		+12,369.62
F. Total Cash Inflow After Taxes		$50,780.82

Figure 5:26 First year cash inflow (componentizing)

15-year life 10% rate		10-year life 15% rate		Total	
Depreciation	Book Value	Depreciation	Book Value	Depreciation	Book Value
	$212,211.21		$84,844.70		$1,000,000.00
$21,221.12	190,990.09	$12,726.70	72,118.00	$77,924.06	922,075.94
19,099.01	171,891.08	10,817.70	61,300.30	70,967.64	851,108.30
17,189.11	154,701.97	9,195.04	52,105.26	64,711.34	786,396.96
15,470.20	139,231.77	7,815.79	44,289.47	59,076.76	727,320.20
13,923.18	125,308.59	6,643.42	37,646.05	53,995.32	673,324.88
12,530.86	112,777.73	5,646.91	31,999.14	49,406.34	623,918.54
11,277.77	101,499.96	4,799.87	27,199.29	45,256.70	578,611.84
99,031.90	113,179.31	59,391.29	25,453.41		

	50-year life 3% rate		25-year life 6% rate		20-year life 7.5% rate	
Year	Depreciation	Book Value	Depreciation	Book Value	Depreciation	Book Value
0		$112,424.52		$245,696.33		$344,823.24
1	$3,372.72	109,051.80	$14,741.78	230,954.55	$25,861.74	318,961.50
2	3,271.55	105,780.25	13,857.27	217,097.28	23,922.11	295,039.39
3	3,173.40	102,606.85	13,025.84	204,071.44	22,127.95	272,911.44
4	3,078.21	99,528.64	12,244.20	191,827.24	20,468.36	252,443.08
5	2,985.86	96,542.78	11,509.63	180,317.61	18,933.23	233,509.85
6	2,896.28	93,646.50	10,819.05	169,498.56	17,513.24	215,996.61
7	2,809.40	90,837.10	10.169.91	159,328.65	16,199.75	199,796.86
Straight Line	15,739.43	96,685.09	68,794.97	176,901.36	120,688.13	224,135.11

Figure 5:27 Depreciation using componentizing

By summing the first-year depreciation amounts for each group, $77,924.06 is determined. This compares to $37,500 in Figure 5.16 for the noncomponentized building. The dollar amount of depreciation is also greater for each year 2 through 7 for the componentized analysis relative to the noncomponentized analysis.

Because the depreciation amounts are increased, the total cash inflow for each of the 7 years as calculated in Figure 5.28 and summarized in Figure 5.30 is greater than the amounts calculated in the noncomponentized analysis. It is noteworthy that the increased depreciation from componentizing actually results in a net taxable loss in years 1 through 5. This results in a positive tax contribution instead of a tax liability for these years.

The accelerated depreciation relating to componentizing does result in

Office Building Example (Using Componentizing)

End of Year	Operating Expenses	Depreciation	Interest	Net Operating Gain	Tax	Total Cash Inflow
1	$100,000.00	$77,924.06	$132,000.00	−$30,924.06	−$12,369.62	$50,780.82
2	100,000.00	70,967.64	130,711.68	− 22,679.32	− 9,071.74	47,482.94
3	100,000.00	64,711.34	129,230.11	− 14,941.45	− 5,976.58	44,387.78
4	100,000.00	59,076.76	127,526.30	− 7,603.06	− 3,041.22	41,452.42
5	100,000.00	53,995.32	125,566.93	− 562.25	− 224.90	38,636.10
6	100,000.00	49,406.34	123,313.65	6,280.01	2,512.00	35,899.20
7	100,000.00	45,256.70	120,722.33	13,020.97	5,208.39	33,202.81

Figure 5:28 Cash inflows for years 1–7 (componentizing)

a larger taxable gain when the building is sold than in the non-componentized analysis. The taxable gain of $441,719.56 from the sale is calculated in Figure 5.29. As was true of the noncomponentized analysis, part of this gain is taxed as ordinary income (the excess accelerated depreciation), and the remaining amount is taxed at the capital gain rate.

A. Cash for Sale
  1. Sale price (1% appreciation) $1,179,348.80
  2. Less 5% commission   58,967.40
  3. Gross cash $1,120,381.40
  4. Less mortgage balance   784,949.40
  5. Adjusted cash income $ 335,432.00

B. Taxes to Be Paid
  1. Gross cash $1,120,381.40
  2. Less book value   678,661.84
  3. Taxable gain $ 441,719.56
  4. Taxable as ordinary income   57,692.44
  5. Taxable as capital gain   384,027.12
  6. Total tax due   84,521.32

C. Net Cash from Sale
  1. Adjusted cash income $ 335,432.00
  2. Less taxes due   84,521.32
  3. Net cash $ 250,910.68

Figure 5:29 Cash flow from sale (componentizing)

Feasibility Estimates and Tax Analysis

The cash inflows and outflows for the componentized feasibility estimate are shown on the time scale in Figure 5.30. Other than the cash inflow from the sale of the building, the annual cash inflows from the componentized analysis are greater than the noncomponentized analysis.

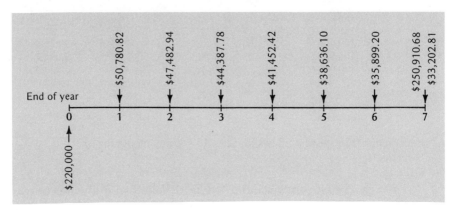

Figure 5:30 Time scale of cash inflows and outflows for componentized $220,000 investment

Year	Cash Inflows	SPPW $i = 20\%$, $n = x$	Discounted Value
1	$ 50,780.82	0.8333	$ 42,315.66
2	47,482.94	0.6944	32,972.15
3	44,387.78	0.5787	25,687.21
4	41,452.42	0.4823	19,992.50
5	38,636.10	0.4019	15,527.85
6	35,899.20	0.3349	12,022.64
7	284,113.49	0.2791	79,296.08
			$227,814.09

Figure 5:31 Discounted cash flows at 20 percent (componentized)

Because the cash inflows in the early years from the componentized analysis are greater than those for the noncomponentized analysis, it follows that the rate of return for the componentized building is greater. The calculation of the rate of return is made in the same manner as in the previous section.

Since a higher rate of return is expected relative to the noncomponentized analysis (in which a 17 percent return was calculated), a guess of 20 percent is used in Figure 5.31. The analysis yields a value of the future cash

Office Building Example (Using Componentizing)

inflows equal to $227,814.09. Because this dollar amount is greater than the building owner's investment of $220,000, a rate of return greater than 20 percent is indicated.

A rate of return of 25 percent is calculated in Figure 5.32. The analysis indicates that this rate of return is too high. Finally, a 22 percent rate of return is calculated in Figure 5.33. The analysis indicates that this is approximately the building's rate of return.

Year	Cash Inflows	SPPW $i = 25\%$, $n = x$	Discounted Value
1	$ 50,780.82	0.8000	$ 40,624.66
2	47,482.94	0.6400	30,389.08
3	44,387.78	0.5120	23,726.54
4	41,452.42	0.3277	13,583.96
5	38,636.10	0.2621	10,126.52
6	35,899.20	0.2097	7,528.06
7	284,113.49	0.1678	47,674.24
			$173,653.06

Figure 5:32 Discounted cash flows at 25 percent (componentized)

Year	Cash Inflows	SPPW $i = 22\%$ $n = x$	Discounted Value
1	$ 50,780.02	.8196	$ 41,619.96
2	47,482.94	.6719	31,903.79
3	44,387.78	.5507	24,444.35
4	41,452.42	.4514	18,711.62
5	38,636.10	.3700	14,295.36
6	35,899.20	.3033	10,888.23
7	284,113.49	.2486	70,630.61
			$212,493.93

Figure 5:33 Discounted cash inflows at 22 percent (componentized)

The componentized depreciation analysis has resulted in an approximately 5 percent increase in the rate of return relative to the noncomponentized analysis. This is a significant difference. In fact, the additional rate of return might be the difference between a good and bad project.

EXERCISE 5.1

FEASIBILITY ESTIMATE

A company builds a new plant for $1.5 million (land and building). The building itself is valued at $1.2 million. The company invests $300,000 cash and obtains a mortgage for $1.2 million at 10 percent interest and payable in 20 end-of-year payments. The firm expects revenue during the first year (assume it is at the end of the year) of $250,000 and expects maintenance and repair (independent of depreciation) of $100,000 (assume it is at the end of the year). The building is to be depreciated over 25 years. Using the most favorable depreciation method for the building in question, determine the after-tax cash flow for the firm for the end of year 1. Assume the company is in a 40 percent tax bracket. The capital recovery factor for an interest rate of 10 percent and 20 periods is 0.11746. Ignore investment tax credit.

EXERCISE 5.2

COMPONENTIZING A BUILDING

Listed below are component parts of a plumbing contract along with the estimated cost of each of the components. Based on your knowledge of plumbing construction, estimate the useful life of each of the components and calculate the first-year depreciation using your calculated lives relative to depreciating the entire plumbing contract with a 40-year life. Assume the construction work is eligible for a 150 percent depreciable life.

Exercise

Component	Cost	Lives
Storm Drains	$ 1,000	
Building Structure		
Roof Drains	8,000	
Service Equipment		
Sewage	8,000	
Plumbing fixtures	10,000	
Plumbing connections	14,000	
Water mains	18,000	
Natural gas mains	6,000	
Heat exchanger	2,000	
Air conditioning connector	2,000	
Processing Facilities		
Industrial waste system	60,000	
Gas process piping	300,000	
Reflection pool piping	5,000	
Steam piping	90,000	
Steam boiler	30,000	
Process pumps	42,000	
Laboratory sink	1,000	
Drinking facilities	1,000	
Emergency shower wash	1,000	
Fire hose cabinet	1,000	
	$600,000	

6
CM Estimates

THE IMPORTANCE OF CM ESTIMATES

As part of its services, the construction manager is often called upon to give the owner a preliminary cost estimate of the proposed project. This estimate may be revised throughout the design phase as materials and aesthetics are determined. These preliminary cost estimates are referred to as CM *estimates* in this chapter.

The purpose of the CM estimate can vary depending on the owner's demands and the type and size of the project. Frequently the estimate is used to support the feasibility estimate (discussed in the previous chapter), to evaluate possible design modifications to keep the project within the owner's budget, to evaluate contractor bids, and as an aid in budgeting cash-flow needs throughout the project.

Estimating and cost control are two related management functions. Of the four project management functions of planning, organizing, directing, and controlling, estimating is considered a planning function, whereas cost control is viewed as a control function. However, the estimate should provide the base for carrying out the control function. In addition, as will be discussed in chapter 10, the control function can be carried out such that data collected as part of this function can serve the needs of future estimates.

Although there is benefit to be gained from an accurate CM estimate, factors such as the lack of detailed drawings, lack of thorough knowledge of construction methods, and limitations on the fees restrict the attainable accuracy. Although it is not expected that the CM estimate be precise, it should be reasonably accurate because one of the justifications for using the CM process is the fact that the CM firm has construction estimating expertise. In chapter 1 it was noted that one of the reasons for the existence of the CM process is the failure of the potential project owner to receive accurate preconstruction estimates from the architect/engineer design team when using the non-CM process. The design profession typically lacks the everyday construction knowledge that enables accurate preconstruction estimates.

The fact that the construction manager is asked to make cost estimates before project drawings and contract drawings are complete means that these estimates are less accurate than the subsequent cost estimates prepared by contractors. It also follows that the procedures and source documents used by the construction manager in preparing preconstruction estimates differ from those used by the construction contractor. Often the use of a given set of procedures for estimating the cost of a project has resulted in the procedures and the related source documents being identified as a specific estimating technique or approach. For example, parameter estimating, factor estimating, and range estimating are CM estimating techniques that are characterized by a specific process and procedure for preparing a preconstruction cost estimate. These and other CM estimating techniques are discussed in this chapter.

CM ESTIMATES: WHEN AND WHY

The CM preliminary cost estimate is made without working drawings or detailed specifications. In fact, the CM may have to make such an estimate from rough design sketches, without dimensions or details, and from an outline specification and a schedule of the owner's space requirements. Recent developments in the construction process have focused attention on the need for accurate preconstruction estimates. The increasing complexity of construction projects, their rapidly increasing costs, and the owner's difficulty with the high cost of obtaining funds all draw attention to the importance of the CM estimate. The use of fast-track or phased construction increases the need for an accurate CM estimate. Should the CM firm fail to provide a reliable estimate, an owner may start a project only to find that there is no capital to complete it.

The types of skills needed by the CM firm for reliable preliminary estimates are many. Included are the following:

1. Knowledge of construction materials
2. Understanding of building design
3. Ability to conceive design details
4. Knowledge of construction trades
5. Acquaintance with construction labor productivity

Although this list is not all inclusive, it is fundamental to the determination of an accurate estimate.

The achievement of an accurate preliminary estimate is tempered by the unavailability of sufficient design detail and the time and fee restrictions on the CM firm. A CM preliminary cost estimate that is within 5 percent of the actual construction cost should be looked upon with favor. Ignoring the insufficient source documents available to the CM firm at the time of the preliminary estimate, the more time the CM firm allocates to the preliminary cost estimate, the more accurate the estimate will tend to be. However, the CM firm must operate within its budget—time is an overhead cost to the owner. The preparation of preconstruction cost estimates is only one of the many CM services the firm must perform within its contract budget.

The marginal increase in the accuracy of the preliminary cost estimate as a function of time spent determining the estimate is difficult to determine. This difficulty does not remove the relevance of the accuracy versus time trade-off. Figure 6.1 indicates the relative accuracy associated with various types of estimates—each of which characteristically requires a different amount of estimating hours. Other than the detailed contractor's estimate, the other estimates can be considered CM cost estimates.

	Expected Accuracy (%)	Approximate Manhours[a]
Unit cost per square foot estimate	40	1
Trade estimate	30	2
Factor estimates	20	5
Parameter estimate	15	10
Contractor detailed estimate	7	150

[a] Assumes a $2 million project.

Figure 6:1   Relative accuracy of estimates

It is difficult to set a number of hours for the determination of the CM preliminary estimate. Instead, the CM firm has to keep in mind the objectives of its estimate, the accuracy and completeness of drawings, and the relative marginal accuracy and benefits obtained for additional estimating time. These factors, along with the owner's willingness to pay for the CM firm's estimating time, should dictate the time expended for the CM preliminary estimate.

The CM preliminary cost estimate can serve several purposes, including the following:

1. Supplement or serve as the owner's feasibility estimate
2. Aid the A/E in designing to a specific budget
3. Assist in the establishment of the owner's funding requirements
4. Serve as a means of evaluating contractor bids
5. Provide the basis for determining progress payments to be made to contractors

Normally a single CM preliminary cost estimate serves several of these purposes. However, should the estimate be prepared to serve a single purpose, the amount of detail and the time at which the estimate is prepared can vary. For example, if an estimate is to be made only to monitor progress payments, the estimate may be made as late as the start of construction. Because actual payments are keyed to the estimate, this CM estimate will likely have to be more accurate than if it had been prepared to serve as a feasibility estimate.

The CM preliminary cost estimate often varies in form and scope depending on whether the project is a building or heavy and highway construction. Building construction consists of projects such as office buildings, industrial plants, commercial buildings, and manufacturing plants. Heavy and highway construction projects include roadways, sewage treatment plants, and bridges.

The project drawings and specifications prepared by the A/E for building construction projects typically leave out the quantities of work the contractor is to perform. In general, the CM preliminary cost estimate does not address the specific work items or quantities. In contrast, drawings for heavy and highway projects typically have the quantities of the various project materials set out. This is especially true in regard to roadway or bridge construction. The quantities will often appear as part of the drawings.

The reasons for setting out the quantities for heavy and highway projects are several. Undoubtedly it is partly due to the repetitive nature of heavy and highway work components. This leads to fewer work items and materials relative to building construction and facilitates the A/E's setting out of the quantities. In addition, there is less variability in the way the contractor can break out work packages and in the technology available to perform heavy and highway work. This in itself makes the quantities more useful to the contractor in estimating.

Contrasted to the CM preliminary cost estimate, the CM firm may also be called upon to perform a *progress estimate* or a *final estimate*. The CM firm's involvement in such estimates is dependent upon the owner's needs and the type of contract used in the construction management agreement with the CM firm. Normally the CM firm would have the responsibility to prepare these estimates.

The progress estimate is the inspection and determination of work completed at the job site. Such an estimate is required periodically so that progress payments to the contractor can be established. It is common for progress estimates to be made monthly.

The final estimate is merely the last progress estimate. It serves the purpose of verifying that all work set out in the contract documents is in fact completed. If this is the case, then and only then is final payment to the contractor authorized. The final estimate can also serve the purpose of determining final quantities of work performed should the final quantities not be set out in the contract documents. For example, the final estimate may be necessary in order to determine the exact length of piling driven by the contractor for which the contractor is to be paid.

UNIT COST ESTIMATES

Unit cost estimates are the quickest and easiest preconstruction estimates for the CM firm to prepare. If based on reliable data, and if used properly, the unit cost estimate provides the potential for an accurate estimate within an acceptable time allowance. Unit cost estimates vary as a function of the base element considered. Perhaps the broadest base unit cost estimate is the *function estimate*. The function estimate measures the cost of a building relative to its use or function. The following are examples of function estimates.

Type of project	Function estimate
School building	Cost per student
Hospital	Cost per bed
Theater	Cost per seat
Parking deck	Cost per parking space

In establishing a preliminary cost estimate by means of a function estimate, the CM firm would be expected to know or approximate the cost per functional element. Multiplying this cost times the quantity of the functional element would yield the total cost estimate.

Perhaps the most commonly used unit cost estimate is the *cost per square foot estimate*. The cost per square foot estimate is based on historical cost per square foot data. By multiplying the historical square foot cost times the calculated square feet of floor area for the proposed building, a CM firm preliminary cost estimate for the building can be determined.

The reason for the acceptance of the unit cost per square foot estimate is that many building costs are very closely related to the square feet of floor area. As indicated in the previous chapter, unit cost per square foot estimates are also widely used when preparing feasibility estimates.

Unit cost per square foot data are readily available in industry cost books. Data from one of these costs books are shown in Figure 6.2. It is common to list the cost per square foot for building type and different qualities of construction. That is, the cost per square foot for a hospital may be shown assuming cheap construction, average construction, or high-quality construction. The purpose of listing several qualities of construction is to enable the CM firm to more accurately estimate the cost of the project in question. Obviously, in order to use such data the CM firm must first determine what quality of construction the project is to have.

Similar to the cost per square foot estimate is the *cost per cubic foot estimate*. The cost per cubic foot estimate relates the cost of a building to

Unit Cost Estimates

Type of Building	Cost per Square Foot
Apartments	$34.50
Banks	62.40
Churches	38.40
Department stores	29.80
Dormitories	45.60
Factories	28.40
Hospitals	68.30
Office buildings	44.50
Schools	39.60
Shopping centers	31.40
Warehouses	25.20

Figure 6:2 Example of unit cost per square foot cost data

its volume. Historical data are collected regarding the cost as a function of the enclosed volume of the building. For example, if the volume of the proposed building is 50,000 cubic feet, this number would be multiplied by a cost per cubic foot as determined from previous similar projects. Cost per cubic foot estimates are usually rather unreliable unless virtually identical buildings are compared. There is not much relationship between the volume of the building and its cost.

Cost per cubic foot estimates are primarily used for structures such as warehouses. Because such structures have widely varying floor heights, the cost per square foot estimates become unreliable. Cost per cubic foot estimates on these types of buildings are more accurate.

Yet another type of unit cost estimate is the *cost per enclosure area estimate*. This unit cost estimate is based on the area of all the horizontal and vertical planes of the building. The interior area of the floors or decks is added to the interior areas of the walls. Using this as a base, this area is divided into the total cost of the building to develop a cost per enclosure area. For similar projects, these data can be used to determine the preliminary cost estimate. The cost per enclosure area estimate is usually reliable because the costs of the building are related to the interior horizontal and vertical planes.

Going one step further in detail is the cost per element or *trade unit cost estimate*. Such an estimate breaks down the total building into basic parts or trades. Past unit cost data of these parts are structured to be used for future preliminary cost estimates. An example of such historical cost data for the cost per element or trade estimate is shown in Figure 6.3. Such an estimate has a potential for accuracy not obtainable from the functional or cost per square foot estimates.

The CM firm may also determine a preliminary cost estimate by relating the quantities of work to each of the sixteen uniform code divisions of work.

Excavation	$ 34,500
Formed concrete	70,000
Exterior masonry	170,500
Interior masonry	64,700
Structural steel	152,300
Ornamental metal	13,600
Carpentry	10,500
Waterproofing	7,400
Roofing	21,400
Doors	16,200
Windows	10,900
Curtain walls	24,800
Drywall	42,300
Plumbing	44,200
Sprinklers	44,000
HAVC	58,200
Electrical	39,800
	$825,300

Figure 6:3 Trade estimate

An example of such an estimate is shown in Figure 6.4 and is often used by the A/E because the A/E firm uses the divisions of work for writing the specifications. In addition, this division of work category is compatible with other architects and engineers.

Division	Work	Estimate
1	General requirements	$ 40,000
2	Site work	30,000
3	Concrete	145,000
4	Masonry	85,500
5	Metals	31,200
6	Wood & plastics	52,400
7	Moisture protection	11,400
8	Doors, windows, & glass	8,600
9	Finishes	32,400
10	Specialties	18,300
11	Equipment	38,500
12	Furnishings	9,200
13	Special construction	19,800
14	Conveying systems	14,800
15	Mechanical	63,200
16	Electrical	21,200
		$621,500

Figure 6:4 Sixteen-division estimate

If used properly, the unit cost estimates described can satisfy the purpose of the preliminary cost estimate. However, the unit cost estimate remains an approximate estimate. The CM firm has to recognize that numerous factors can change the cost of a proposed building. Some of these factors are

1. Changes in quality of construction
2. Changes in aesthetics or design
3. Site conditions of the project
4. Weather conditions affecting the project
5. The skills of the labor force
6. The morale of the labor force
7. The material prices and wage rates
8. The experience and skill of the project management team

Other factors also can change the cost of the proposed project. However, given the time constraint usually placed on the preliminary estimate, the unit cost estimate serves well as a first cost estimate. It should be remembered that it is difficult to obtain total accuracy if all input data are not known. As such, the unit cost estimate may in fact be the most accurate given the time at which it is made.

PARAMETER ESTIMATING

A recently evolving CM preliminary cost estimate method is the *parameter estimate*. Parameter estimating can be used by the CM firm to give the owner an approximate cost of a project and also to enable the CM firm to evaluate contractor bids.

Engineering News Record first published parameter cost estimates in 1966. From that time the periodical has expanded the concept of parameter estimating. Whereas only the gross floor area was used as a parameter in early years of parameter estimating, it has now been expanded to include several parameter measures. For *Engineering News Record*'s parameter cost breakouts, contractors, owners, designers, or construction managers allocate lump-sum costs to trades or component systems of a building's construction. Parameter measures are then chosen to divide into the trade cost, one parameter to a trade. The choice relates the parameter measure to the function of the trade. For example, structural steel cost may be related to the gross area supported, and dry wall cost to interior area. Trades such as acoustical ceilings would be attributed to net finished area. The parameter costs can yield guidelines to estimate the cost of a similar project or to judge the validity of contractors' quotes.

Let us look at an example of collected parameter cost data. This data could easily be collected by the CM firm or could be taken from published cost data such as in *Engineering News Record*. Parameter cost data for an

Type of building	Office
Location	Troy, Mich.
Construction start/complete	Aug./Oct.
Type of owner	Private
Frame	Structural steel
Exterior walls	Metal curtain wall
Special site work	None
Fire rating	2-A

PARAMETER MEASURES

1. Gross enclosed floor area	650,000 sf
2. Gross area supported (excluding slab on grade)	608,400 sf
3. Total basement floor area	41,600 sf
4. Roof area	70,000 sf
5. Net finished area	500,000 sf
6. Number of floors including basements	26
7. Number of floors excluding basements	25
8. Area of face brick	0
9. Area of other exterior wall	150,000 sf
10. Area of curtain wall including glass	221,000 sf
11. Store-front perimeter	300 sf
12. Interior partitions	30,000 sf
13. HVAC	1,700 tons
14. Parking area	890,000 sf

OTHER MEASURES

Area of typical floor	19,400 sf
Story height, typical floor	152 in
Lobby area	2,000 sf
Number of plumbing fixtures	460
Number of elevators	12

DESIGN RATIOS:

A/C ton per building sq. ft.	0.0026
Parking square feet per building	1.3692

Figure 6:5 Parameter data

office building are shown in Figure 6.5. It contains data regarding the type of project, parameter measures, and trade data. Other data shown, but less frequently used, are other measures of parameters and design ratios. The descriptive information regarding the type of project is used merely to classify similar types of projects. That is, should the CM firm attempt to estimate an office building, it would attempt to match the future project with parameter cost data for a similar type of project. Information such as the type of frame, the exterior walls, and the design ratios would aid the CM firm in selecting a similar type of project from its parameter cost library.

Parameter Estimating

Area or Trade	Code	Parameter Cost Unit	Cost	Total Cost $	%
General conditions and fee	1	sf	2.19	1,425,000	9.07
Sitework (clearing and grubbing)	1	sf	0.35	226,000	1.44
Excavation	3	sf	4.57	190,000	1.21
Foundation	2	sf	0.67	410,000	2.61
Formed concrete	2	sf	2.71	1,650,000	10.50
Interior masonry	12	lf	6.20	186,000	1.18
Structural steel	2	sf	1.23	750,000	4.77
Miscellaneous metal, including stairs	2	sf	0.38	230,000	1.46
Carpentry	5	sf	0.18	90,000	0.57
Waterproofing and dampproofing	10	sf	0.10	22,000	0.14
Roofing and flashing	4	sf	1.64	115,000	0.73
Metal doors, frames, windows	5	sf	0.11	55,000	0.35
Hardware	5	sf	0.12	60,000	0.38
Glazing	11	lf	333.33	100,000	0.64
Curtain wall	10	sf	3.96	875,000	5.57
Lath and plaster	5	sf	0.08	40,000	0.25
Dry wall	12	lf	28.90	867,000	5.52
Tile work	5	sf	0.10	49,000	0.31
Accoustical ceiling	5	sf	0.18	90,000	0.57
Resilient flooring	5	sf	0.04	18,000	0.11
Carpet	5	sf	0.03	17,000	0.11
Painting	5	sf	0.17	87,000	0.55
Toilet partitions	5	sf	0.03	13,000	0.08
Elevators	om	ea	70,000	70,000	0.44
Plumbing	1	sf	5.15	3,350,000	21.35
HVAC	1	sf	5.15	3,350,000	21.35
Electrical	1	sf	1.77	1,150,000	7.32
Parking, paved	14	sf	0.25	223,000	1.42
TOTAL	1	sf	24.17	15,708,000	100.00

Figure 6:5 *continued*

The parameter measures shown in Figure 6.5 indicate the actual measures for a previously constructed office building. That is, the project has a gross enclosed floor area of 650,000 square feet and the frame is structural steel. Other parameter measures are shown in the figure. These measured parameters will be used to relate the actual cost of the project by segments or trades. Let us look at the data regarding the trades. In the first column a code is indicated. This code relates the type of work in question to one of the parameters. For example, the general conditions and fee for the office building are indicated to be a measure or function of the parameter of gross enclosed floor area. On the other hand, the interior masonry shown in the trade has a code of 12. This indicates that the interior masonry is most a function of the linear feet or partition wall. The measured linear feet of partition walls was 30,000 Dividing this quantity into the total cost amount develops a parameter cost of $6.20. Similar parameter unit costs are developed and shown in Figure 6.5. Another column relates the percentage of trade cost amount to the total cost of the project. For example, the interior masonry is shown as 1.18 percent. This is developed by dividing the total costs for the interior masonry, $186,000, by the total cost of the project of $15,708,000.

The value of the parameter cost data shown in Figure 6.5 is that in determining a measure of preliminary cost, the CM firm merely has to take off the parameter measure quantities and have a historical library of parameter costs by trade. For the office building this would mean that the CM firm would only have to take off fourteen measures of project quantities. This compares to several hundred quantities that are characteristic of a contractor's detailed estimate. The unit cost estimate is in reality a one-parameter estimate. Typically the parameter estimate used by the CM firm would be restricted to eight to fifteen parameters.

Let us now look at how a parameter estimate can be made from previous parameter cost data. Let us assume we are faced with determining a preliminary cost estimate for another office project. Let us further assume that the office building in question is similar in quality and construction to the office building for which we have collected data in Figure 6.5. The type of project is described in Figure 6.6. The parameter measures are shown; for example, the office building we wish to estimate has gross enclosed floor area of 395,000 square feet. Other parameter measures are shown in the figure. The parameter cost data for the trades as determined from the previous office building project are also shown. In reality, the parameter cost data we would use for estimating would likely be an accumulation of several past projects. However, for purposes of demonstration, let us assume that these accumulated cost data are merely those shown in the previous example. This cost data is needed to determine the estimated cost of the project in question. The unit parameter cost for general conditions for completing a project was determined as $2.19. Because we identified this type of construction as being related to gross enclosed floor area we will now multiply the area for the project by the unit parameter cost, that is, $2.19 times 395,000 square feet. This would give us the estimated general conditions and fee cost for the proposed office building project.

Similar trade costs for the other type of work proposed for the project in question would be determined in a similar manner. For example, from previous parameter cost data we have determined a unit parameter cost for interior masonry of $6.20 per linear foot. The new project has an estimated linear footage of partition walls of 35,500. Because the interior masonry cost

Parameter Measures

1. Gross enclosed floor area — 395,000 sf
2. Gross area supported (excluding slab on grade) — 350,000 sf
3. Total basement floor area — 45,000 sf
4. Roof area — 40,000 sf
5. Net finished area — 320,000 sf
6. Number of floors including basements — 12
7. Number of floors excluding basements — 11
8. Area of face brick — 0
9. Area of other exterior wall — 80,000 sf
10. Area of curtain wall including glass — 110,000 sf
11. Store-front perimeter — 300 sf
12. Interior partitions — 35,500 lf
13. HVAC — 1,200 tons
14. Parking area — 200,000 sf

Trade	Parameter Cost Code	Unit	Cost	Total Cost Amount	%
General conditions and fee	1	sf	2.19	$ 865,050	8.62
Sitework (clearing and grubbing)	1	sf	0.35	138,250	1.38
Excavation	3	sf	4.57	205,650	2.05
Foundation	2	sf	0.67	234,500	2.34
Formed concrete	2	sf	2.71	948,500	9.45
Interior masonry	12	lf	6.20	220,100	2.20
Structural steel	2	sf	1.23	430,500	4.29
Miscellaneous metal, including stairs	2	sf	0.38	133,000	1.33
Carpentry	5	sf	0.18	57,600	0.57
Waterproofing and dampproofing	10	sf	0.10	11,000	0.11
Roofing and flashing	4	sf	1.64	65,600	0.65
Metal doors, frames, windows	5	sf	0.11	35,200	0.35
Hardware	5	sf	0.12	38,400	0.38
Glazing	11	lf	333.33	100,000	1.00
Curtain wall	10	sf	3.96	435,600	4.34
Lath and plaster	5	sf	0.08	25,600	0.25
Drywall	12	lf	28.90	1,025,950	10.22
Tile work	5	sf	0.10	32,000	0.32
Accoustical ceiling	5	sf	0.18	57,600	0.57
Resilient flooring	5	sf	0.04	12,800	0.13
Carpet	5	sf	0.03	9,600	0.10
Painting	5	sf	0.17	54,400	0.54
Toilet partitions	5	sf	0.03	9,600	0.10
Elevators	om	ea	70,000	70,000	0.70
Plumbing	1	sf	5.15	2,034,250	20.27
HVAC	1	sf	5.15	2,034,250	20.27
Electrical	1	sf	1.77	699,150	6.97
Parking, paved	14	sf	0.25	50,000	0.50
TOTAL	1	sf	25.40	10,034,150	100.00

Figure 6:6 Parameter data for a proposed office building

is identified as a function of linear feet of partition walls we would multiply the 35,500 by $6.20 to determine the total estimated masonry cost of $220,100. The column of percentages is determined by simply taking the estimated cost of a given trade and dividing it by the calculated sum of the total estimated cost. In this manner we determine the total estimated cost of the proposed project and the components of the cost identified by trade. The result is that the CM can determine an approximate cost of the project and also develop component trade costs that can be used to evaluate contractor bids.

FACTOR ESTIMATING

Factor estimating can be used by the CM firm to make preliminary cost estimates for various types of projects. However, the factor estimate is best used for projects with a predominant cost component. Often this component is the purchased equipment for the building. Examples of such projects are oil refineries and foundries. These and other types of buildings are referred to as *process plants*.

The factor estimate is also based on historical data. However, unlike other estimating models that gather and structure past cost data, the factor estimate develops factors for each component as a function of a predominant cost.

Type of Work	Cost	Factors
General Conditions	$ 90,000	0.09
Excavation	70,000	0.07
Framing System	220,000	0.22
Equipment	1,000,000	1.00
Equipment Installation	180,000	0.18
Process Piping	700,000	0.70
Instrumentation Costs	200,000	0.20
Finish Material	150,000	0.15
Electrical	100,000	0.10
Plumbing	180,000	0.18
Mechanical	440,000	0.44
Total Project Cost	$3,330,000	

Figure 6:7 Factor estimate data from a past project

Figure 6.7 shows collected factor cost data. It can be observed that the total project is broken up into a series of components. Example components are the equipment costs, the equipment installation cost, the process piping, the instrumentation cost, and so on. After listing the items or components of

Factor Estimating

work, a factor for each is listed. In this factor estimate the purchased equipment cost is given a factor of 1.00. The factors for such items as the equipment installation are developed by observing the actual cost as determined from previous projects or the actual construction costs. For example, the process piping cost for the project in question is $700,000. The purchased equipment cost was $1 million. Thus the process piping cost was 70 percent of the purchased equipment cost or 0.70. The other factors shown are calculated in the same manner. That is, the actual cost is divided by the purchased equipment cost to determine the factor for that particular component.

By collecting data for numerous similar projects, the reliability of the factors becomes more reliable. That is, if the data represent a rather wide assortment of different project conditions, then we can be fairly assured that the process piping cost will be 70 percent of the actual purchased equipment cost. The data that are kept for future estimating are the factors and not the actual costs from past projects. It is these factors that are used to aid the CM firm in future estimating projects.

The theory behind factor estimating is that components of a given type of project will have the same relative cost as a function of a key or predominant cost for each and every project. For example, for a steel mill, the processing equipment often dictates the cost of the building components. It should be noted also that component factors can be greater than one. This is because the purchased equipment cost, assuming it is the key cost, may not be in excess of every other component cost for that type of project.

The collected factor cost data can be used to determine a preliminary cost estimate for a project. Let us assume that we are building a project similar to the one in Figure 6.7. However, this project will not necessarily be of the same magnitude. In particular, let us assume that we estimate from bids of equipment from vendors that the purchased equipment cost for the project we are proposing is $600,000. Having developed the factors shown in the previous example, we can now use these factors to estimate the component cost of the project we are proposing. This is done in Figure 6.8.

Any one component cost of the proposed project is determined by multiplying the historical factor for that component by the estimated purchased equipment cost for the proposed project. For example, the electrical factor from the previous project was 10 percent or 0.10. As such, the estimated cost of electrical work for the proposed project is 10 percent times the purchased equipment cost of $600,000. Thus the electrical cost is estimated at $60,000. The other cost components of the project are determined in a similar manner. Summing all of the estimated costs yields an estimated project cost. Assuming that the collected factors are reliable and that the components have the same cost relationship to the predominant cost as in previous projects, then the estimate is fairly reliable. However, like other preliminary cost estimates, the factor estimate is only intended to be approximate.

Type of Work	Factor	Projected Cost
General conditions	0.09	$ 54,000
Excavation	0.07	42,000
Framing system	0.22	132,000
Equipment	1.00	600,000
Equipment installation	0.18	108,000
Process piping	0.70	420,000
Instrumentation costs	0.20	120,000
Finish material	0.15	90,000
Electrical	0.10	60,000
Plumbing	0.18	108,000
Mechanical	0.44	264,000
Total Project Cost		$1,998,000

Figure 6:8 Projected building cost using factor estimating

RANGE ESTIMATING

An estimate, by definition, is uncertain. The amount of uncertainty that a cost estimate of a building has is dependent on the skill and knowledge of the preparer as well as the characteristics of the building being proposed and the timing of the estimate. One means of recognizing and evaluating the uncertainty of an estimate is through the use of *range estimating*. Its use has proved an effective means of determining accurate preconstruction estimates.

Range estimating is in part a computer software program. Therefore, the range estimating process cannot be illustrated in total without the presentation of the probability theory that is part of the program. However, because of the growing popularity of range estimating and its effective use by construction managers, an overview of the process is given in this section.

Range estimating has the objective of setting out a range of possible project costs or probabilities of various project costs within this range. In conventional estimating, the estimator ultimately represents the anticipated cost of the project by a single number. This single dollar number is set out even though the estimator knows that hundreds if not thousands of actual project costs are possible. The range estimating process calculates an expected range of total project or building costs as a function of the range of expected costs of the project's individual work phases or packages. The result is that the user of the range estimating process can equate risks with various possible project budgets. For example, a range estimate may indicate that there is an 80 percent probability that a project will cost less than $1 million, an 85 percent probability that it will cost less than $1.1 million, a 90 percent probability that it will cost less than $1.2 million, and so on. Range estimating also sets

Range Estimating

Work Package	Cost
General conditions and fee	$ 165,400
Sitework	26,200
Excavation	22,000
Foundation	47,600
Formed concrete	75,400
Interior masonry	21,600
Structural steel	319,000
Miscellaneous metal	26,600
Carpentry	10,400
Waterproofing and dampproofing	2,600
Roofing and flashing	13,400
Metal doors and frames, windows	6,400
Hardware	7,000
Glazing	11,600
Curtain wall	217,600
Lath and plaster	4,600
Drywall	106,000
Tile work	5,600
Accoustical ceiling	10,400
Resilient flooring	2,000
Carpet	2,000
Painting	10,000
Toilet partitions	1,600
Elevators	10,000
Plumbing	388,600
HVAC	331,200
Electrical	129,400
Parking	25,800
Total	$2,000,000

Figure 6:9 Expected package costs

out critical phases or packages that can greatly affect the project either adversely or favorably.

Let us illustrate the range estimating process by means of an example. Shown in Figure 6.9 is an estimate of a building's cost as might be prepared by an architect/engineer or a construction manager. The estimated cost is $2 million. This amount might be determined with a parameter estimate, a factor estimate, or another estimating process. We might consider the estimated cost of each work package and the total $2 million as target costs.

The range estimating process does not limit itself to an estimate of a single cost for each work package or phase. Instead, the user of the process states a target cost, a lowest estimated cost, a highest estimated cost, and a confidence limit or likelihood that the actual cost will be equal to or less than the target cost. This is illustrated in Figure 6.10. It should be noted that

Work Package	Target Cost	%	Lowest Cost	Highest Cost
General conditions and fee	$ 165,400	65	$ 154,000	$ 186,000
Sitework	26,200	70	24,000	27,000
Excavation	22,000	80	20,500	23,400
Foundation	47,600	40	45,300	57,000
Formed concrete	75,400	80	72,000	75,600
Interior masonry	21,600	75	21,000	27,720
Structural steel	319,000	60	304,000	342,000
Miscellaneous metal	26,600	80	24,400	29,300
Carpentry	10,400	80	9,800	10,400
Waterproofing and dampproofing	2,600	75	2,000	3,200
Roofing and flashing	13,400	60	12,500	17,400
Metal door and frames, windows	6,400	80	6,400	6,400
Hardware	7,000	80	6,400	7,500
Glazing	11,600	80	10,800	15,400
Curtain wall	217,600	90	210,000	242,000
Lath and plaster	4,600	80	4,400	5,500
Drywall	106,000	70	101,000	121,000
Tile work	5,600	65	5,100	6,300
Accoustical ceiling	10,400	70	9,400	12,200
Resilient flooring	2,000	80	1,900	2,200
Carpet	2,000	70	2,000	2,000
Painting	10,000	70	9,400	11,500
Toilet partitions	1,600	60	1,500	2,100
Elevators	10,000	70	9,500	12,000
Plumbing	388,600	80	328,000	346,000
HVAC	331,200	70	315,000	381,200
Electrical	129,400	70	126,000	134,600
Parking	25,800	80	22,000	33,000
TOTAL	$2,000,000		$1,858,300	$2,139,920

Figure 6:10   Range costs

the lowest and highest costs are unlikely, but should be viewed as possible. Although a certain amount of guessing or subjectivity goes into setting out the lowest and highest cost estimates for the individual work packages or phases, an error made in setting these ranges will have less effect on the estimate's accuracy than an error in another estimating process.

The individual work packages or phases illustrated in Figure 6.10 might be compatible with the manner in which individual work packages are subsequently awarded to contractors. If this is done, the CM firm can in effect prepare an estimate that may accommodate the awarding and controlling of individual construction contracts. Once the construction manager has determined its individual work package cost estimates, including the low/high range, the next step is processing the information. Obviously thousands of cost combinations can result when one starts combining the possible costs

Range Estimating

of individual work packages. The calculation and weighting of the possible combinations would be extremely time consuming, if not impossible, if the process were done manually.

As indicated in Figure 6.10, based on a summation of possible lowest costs for the individual packages, the total project cost could be as low as $1,858,300. On the other hand, based on a summation of possible highest costs, the total project could cost $2,139,920. Neither of these costs are likely. They represent extremes that have only minute probabilities associated with them. More likely a total project cost within this range will occur.

The knowledge of the range of project costs and the likelihood of overrunning a single cost helps the construction manager and owner in several ways. For one, it is possible to equate risk. It is also possible to budget for contingencies or to redesign aspects of the proposed project to decrease the potential range of costs.

Additional computer output can also be generated to yield information regarding the criticality of individual work packages or phases. In other words, some work packages affect a possible project cost overrun more adversely than others. For example, the analysis may yield information that indicates that the foundation-related work accounts for 20 percent of a projected possible cost overrun. This information can point the construction manager toward tight control over this phase of the project.

The uses of the cost profile shown in Figure 6.11 and related range estimating output are left to the creativity of the construction manager. Independent of this, the range estimate provides more potential for management control than the single dollar cost estimate illustrated in Figure 6.9.

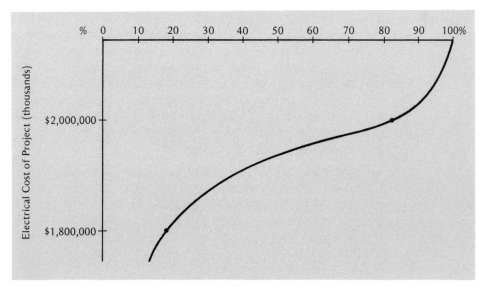

Figure 6:11 Cost profile for range estimating by percent choice of overrun

EXERCISE 6.1

DESIGNING A PARAMETER ESTIMATING SYSTEM

The parameter estimating process depends on the ability to correlate various component or trade costs for a proposed project to various scope parameters that are known early in the design phase. The components or trades and the parameters identified depend in part on the type of building or project being analyzed. Assume a parameter estimating system is to be designed for estimating the cost of a residential single-family home. The objective is to design a system that correlates the various trades necessary to build a house relative to scope parameters normally defined by a potential homeowner at the time it seeks a contractor. Design a parameter estimating system to include 1) a set of parameters, 2) a list of trades, and 3) an identification of a parameter that each trade is related to in terms of estimating its costs.

EXERCISE 6.2

VARIABLES IN PRICING WORK

As a part of the range estimating process, the estimator must set out a possible lowest cost and highest cost as well as target cost for each work component or package. These lowest and highest dollar amounts are meant to be more than merely guesses. Instead, they are to reflect possible cost ranges based on factors that can favorably or adversely affect the target or expected cost. Assuming a building construction project, list factors that must be considered in evaluating the possible ranges of cost for a given work component or package.

7
Value Engineering

DESIGN INEFFICIENCIES

The project owner and layman are often critical of the performance of the contractor team during the construction phase of a project. These individuals may observe daily construction phase inefficiencies at a job site by merely casually observing the construction. Weather delays, delays that result from equipment breakdowns, labor waiting, unavailable material, or delays resulting from jurisdiction disputes between two or more contractors or labor crafts all are visible and well publicized. The result is that the project owner attaches what he or she considers unnecessary costs to the performance of the construction phase of the project.

The project owner and layman are usually less critical of the architectural design team and the contract documents, including the drawings and specifications. Their lack of knowledge of the design process is undoubtedly part of the reason they are less critical of this phase. The fact that the contract documents are prepared at the office of the architect or engineer, away from the close scrutiny of an external party, also leads to a less critical evaluation.

Even though the design contract documents for a project are less critically evaluated, there are often cost, time, and quality inefficiencies associated with this phase. Inefficiencies that can result include delays in completing various phases of the design, thus adding to the overall project time and cost; inferior design in regard to the selection of materials and project systems; and a design that relates to or even promotes inefficiencies in the construction phase.

Part of the CM process includes the CM firm's "policing" the design phase. In addition to inputting a "construction flavor" into the design of a project, the CM firm performs a critical review of this phase with the objective of eliminating or minimizing the above three design phase inefficiencies.

The CM firm does not actually perform the project design. It assists the design firm in this function by adding its construction knowledge and critically reviewing the contract documents as they are prepared. The architect/engineer design team is not in charge of the CM firm or vice versa. The two entities are expected to work together to prepare contract documents that provide a quality project within the time and cost limitations of the owner. Value engineering is a management process that can accomplish these objectives.

Value engineering is a well-publicized CM management technique. Although it is not an all-encompassing tool that ensures optimal project design, value engineering can result in elimination or reduction of significant project quality, time, and cost inefficiencies. Most importantly, value engineering can serve as the communication tool for ensuring that the CM firm does provide the project owner one of its contracted services, a critical review of the project design.

The alternative to value engineering is the reliance on the CM firm's ability to input its knowledge and experience into the design phase in a somewhat subjective manner. Dependence on this unformalized, subjective process may result in the nonperformance of the critical review of the contract documents. It is also possible that the CM firm's use of its "own way" of policing the design phase may result in an adverse relationship between the architect/engineer design team and the CM firm. The CM firm may too harshly criticize the design firm just for the sake of criticizing it. This undoubtedly will not prove effective as to the quality, time, and cost objectives.

The concept of value engineering in the CM process is to provide the means for the CM firm to work with the architect/engineering design firm. In fact, the critical review team of the design phase of a project can include individuals other than the CM firm and the design firm. A project owner's representative, a contractor, a material supplier, and a potential user or maintainer of a proposed project can be effective members of the design review team. The means for all of these individuals to accomplish their common project goal can be value engineering.

Value engineering has several benefits. It is simple in structure and application. This simplicity of application results in ease of use independent of the skills or background of the individual applying it. Benefits of value engineering, the seven steps of performing the value engineering process, and the implementation of value engineering in the CM process are discussed in this chapter.

BENEFITS OF VALUE ENGINEERING

Value engineering should not be viewed as an all-encompassing management technique for optimizing the design phase of a project. Instead, value engineering should be viewed as a technique that enhances the performance of a

Benefits of Value Engineering

critical analysis of a design with the objective of improving it. The improved design can lessen overall project cost, speed up the time of completion, or improve quality.

Like most models or management techniques, value engineering is not a solution in itself. The value engineering process provides a framework by which the construction manager or the designer can work together creatively. In effect, one can view value engineering as a process that promotes the "reinventing of the wheel." The process draws attention to the use of alternative materials to accomplish the same function; the spacial layout of floor space; the alternative types of systems being designed, including structural, electrical, mechanical, and finishes. Numerous alternatives for each of these considerations exist. Each alternative has project cost, duration, and quality implications. The objective of value engineering is to identify the alternatives and to weigh the benefits of each. The mere fact that the alternatives are identified and their relative merits evaluated will enhance the possibility that the resulting design and accompanying contract documents are optimal or near optimal in regard to overall project cost, duration, and quality.

Value engineering has three favorable attributes that enhance its use for critically reviewing the project drawings and related contract documents.

1. Provides a common language for use by individuals from several disciplines
2. Provides a formal step-by-step process
3. Results in a functional estimate that enhances creativity and innovation

As we will illustrate in the next section of this chapter, the value engineering process imposes a "verb-noun" language as part of its implementation. Multiple users of the management technique all use the same language. This eliminates each of the value engineering participants from speaking in the technical language of their professions. The use of a structured common language enhances communication between the CM firm and the project design team. It also becomes possible for the project owner, a material supplier, and the custodian of the planned project to communicate with the CM firm or the design firm and to contribute to their efforts.

Value engineering is a formal process that uses seven steps. These steps are illustrated in the following section. The fact that it is a well-defined process helps remove the subjectivity or vagueness of how the CM firm is to contribute its construction knowledge and its policing role to the design phase of a project. Although one should not discount the experience or creativity of the CM firm in performing these functions, the lack of the use of a well-defined, step-by-step process may result in the CM firm failing to perform the above services or performing them inadequately.

The other favorable attribute of value engineering is the fact that the result of the value engineering exercise, the preparation of a functional estimate, can promote creativity and/or draw attention to design inefficiencies. The project design team and contractor team are more accustomed to "bill-of-material" types of estimates. For example, in preparing an estimate of the cost of construction of a room, the contractor will likely determine the labor and material costs for constructing the wall framing, painting the walls, installing individual fixtures, and laying the carpeting. This can be considered a bill-of-material estimate. The analysis of a bill-of-material estimate inspires little creativity and does not easily center on inefficiencies or wasteful items. For example, the fact that rugs for the room in question cost $1600 draws little or no attention to their cost relative to their worth to the owner.

The value engineering process would also result in the preparation of an estimate of the cost of the room. However, the estimate would be a functional estimate. The costs related to aesthetics, noise insulation, environmental or temperature control, structural integrity, room flexibility, and so on, would be determined. The advocate of value engineering argues that a functional estimate results in a more creative analysis of design and centers the value engineering on design inefficiencies. For example, the fact that the value engineering analysis of a proposed design for a room indicates that 30 percent of the estimated room cost relates to the aesthetics of the room might draw attention to possible cost reduction for this function or the use of more economical materials.

The three attributes of value engineering do not by themselves ensure that the process will result in an improved design. The user of the process must still be creative and innovative when evaluating design alternatives. However, the three attributes do enhance the probability that a critical analysis will be made of the design and also provide the framework for creativity by the user.

THE VALUE ENGINEERING PROCESS— THE SEVEN STEPS

Value engineering is a formalized step-by-step process. Because of the defined seven-step process, value engineering is teachable and more importantly usable by several individuals, each possessing different skills and experiences.

The seven-step value engineering process will now be illustrated by means of an example, the proposed design of a door. Although a door is normally too small a building element to justify a value engineering analysis, it is used because the reader can relate to the functions of a door, its cost elements, and so on. In practice, the value engineering team would focus on larger cost items, including the spacial requirements; various subsystems, such as the structural, electrical and mechanical systems; and so on. However,

The Value Engineering Process—The Seven Steps

even the door example becomes relevant and deserving of a value engineering effort if a building has numerous similar doors as part of its design. For example, in a hotel complex, the savings from a value engineering of a door multiplies several fold if there are a hundred or even a thousand doors in the complex. However, the reader is cautioned to focus on the seven value engineering steps in the following example, rather than on the example itself. Given the knowledge of the seven steps, the reader can apply them to the analysis of more real-life situations.

Step 1—Identify Functions

Step 1 of the value engineering process is for the value engineering team to list the functions of the design element to be analyzed. Specifically, the team has to categorize each function of the building element by means of two words, a verb and a noun. The verb-noun language is the language of the value engineer. The limitation of listing the element's functions with two words forces the users of the process to share a common language. Without this limitation, each team participant would express the building element's functions using their own language, a language that may be foreign to or misinterpreted by the other team members.

Example functions of a building element include "insulate sound," "hold fixtures," "provide security," and so on. Note that functions are not materials or things. They are what the building element is to accomplish for the building owner.

The listing of building-element functions is a brainstorming exercise. Two individuals on the value engineering team may foresee the functions or purposes of a proposed building element differently. Given these differing opinions it is necessary that they be discussed and evaluated with the possibility that what might appear to be a function is not and vice versa. The inclusion of the project owner in the performance of this first value engineering step is recommended. Forcing the owner to make decisions regarding the functions to be served in itself can prove a worthwhile effort.

The value engineering team should consider all possible functions in its performance of step 1. Some possible functions may be deleted after their consideration if they are judged not worthy of the necessary expenditures.

It is true that the value engineering team can on occasion forget to list all of the functions of the building element being analyzed. However, when this occurs it is usually an indication that the omitted functions were of minor importance.

Step 1 of the value engineering of a door is illustrated in Figure 7.1. As noted, the door is a simplistic example. However, the functions listed in the figure are representative of the performance of step 1. The reader's attention is drawn to the use of the verb-noun language.

> Value Engineering: Example Solution (Interior Door)
>
> Step 1. Identify Functions
> 1. Provide access
> 2. Insulate sound
> 3. Facilitate safety
> 4. Provide aesthetics
> 5. Provide security
> 6. Regulate temperature
> 7. Provide durability

Figure 7:1 Value engineering—Step One

Step 2—Initial Estimate Value per Function

The next step in the value engineering process is illustrated in Figure 7.2. In this step the value engineering team establishes the relative worth (on a percentage basis) of each of the functions listed in Figure 7.1. This step can be referred to as the value engineering team's "gut-feeling" or initial estimate.

> Value Engineering: Example Solution (Interior Door)
>
> Step 2. Estimate Value per Function
>
> | Function | Initial Estimate of Value |
> |---|---|
> | 1. Provide access | 45 |
> | 2. Insulate sound | 20 |
> | 3. Facilitate safety | 10 |
> | 4. Provide aesthetics | 10 |
> | 5. Provide security | 5 |
> | 6. Regulate temperature | 5 |
> | 7. Provide durability | 5 |

Figure 7:2 Value engineering—Step Two

No detailed analysis of the building element or its materials or costs is made as part of step 2. Instead, the team uses common sense to determine the initial estimate of function value. Obviously if the potential project owner is part of the value engineering team, his or her statement of the relative worth or value of each function might serve as the basis for the percentages

The Value Engineering Process—The Seven Steps

listed in step 2. However, more likely the team members have some discussion and compromising before the percentages are established.

In setting out the percentages of worth or value in step 2 the value engineering team is setting out how it believes the costs of the design of the building elements should be structured. As part of step 7, this initial evaluation or distribution of worth will be compared to the actual cost distribution as determined in value engineering steps 3 through 6.

The establishment of the initial estimate of functional value may appear very subjective. However, in practice two or more members of the value engineering team will often be very close in their initial distribution of the relative worth of the individual functions. Any differences have to be resolved through the leadership and persuasion of the CM firm.

Step 3—List Parts or Components

The third step of this process is the listing of all of the parts or components that make up the building element being analyzed. In effect, it is the bill of materials that would serve as the means of a contractor preparing a detailed estimate.

The number of parts to be listed in this step is dependent on the building element being analyzed. If an entire room is being analyzed, the number of parts would be many, including wall finishes, rugs, windows, door frame, door, and so on. The number of parts for a door analysis may be limited to those in Figure 7.3.

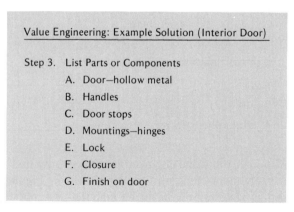

Figure 7:3 Value engineering—Step Three

As a guide to what is and what is not a component, parts than can be purchased separately should be listed separately. For example, a door knob including a lock can be purchased separately from a door and therefore should be listed separately (especially in light of the fact that numerous different types of door knobs and lock combinations could be selected). On the other hand, the screws that are part of the door knob are an integral part of the door knob itself and should not be listed as a separate part.

It is important to note that although value engineering steps 1 and 2 are and should be performed independent of the actual door being analyzed or the actual proposed contract drawings, step 3 addresses the listing of the parts of the specific door or set of drawings being analyzed. In effect, the first two steps are conceptualizing. The functions listed in step 1 are functions of a door, independent of the actual door being proposed. The same is true of step 2. However, in step 3 the actual parts of the proposed door are listed.

Step 4—Cost per Function

Step 4 is to price the individual parts that were listed in step 3, including the material cost, the installation cost, and the sum of both—the total cost. In effect, this step is the same as a contractor would do in preparing a detailed bill-of-material estimate for bidding purposes.

Value Engineering: Example Solution (Interior Door)

Step 4. Cost per Function

Part	$Labor	$Material	$Total
A. Door—hollow metal	30.00	60.00	90.00
B. Handles	14.00	32.00	46.00
C. Door stops	4.00	13.00	17.00
D. Mountings—hinges	14.00	12.00	26.00
E. Lock	10.00	8.00	18.00
F. Closure	14.00	47.00	61.00
G. Finish on door	4.00	4.00	8.00
	$90.00	$176.00	$266.00

Figure 7:4 Value engineering—Step Four

Although the value engineering process only uses the total cost of each part in subsequent steps, the format shown in Figure 7.4, determining the total cost by determining both the material cost and the installation cost, is recommended. The source of the costs of the material and installation costs for the individual parts can come from two different sources. For example, the material supplier on the value engineering team may be best qualified to give an accurate estimate of the raw material cost. On the other hand, a contractor on the team may be most qualified to set out the labor or installation cost. Because the accuracy of the total cost is dependent on the accuracy of both the material and installation costs, it follows that the expertise of both parties

The Value Engineering Process—The Seven Steps

is needed. In fact, a justification for the value engineering process centers around the fact that without recognition of accurate material and installation costs, the evolving design cannot be optimal.

The need to recognize the costs and to ensure that these estimates are accurate is crucial to any subsequent cost reductions imposed on the design. What to cut and what materials can be afforded are dependent on the designer's or value engineering team's interpretation of real cost on a part-by-part basis. The result is that the correct performance of step 4 dictates the validity of subsequent value engineering steps.

Step 5—Identify Function of Each Part

Step 5 is perhaps the hardest and most subjective step in the value engineering process. The step is sometimes referred to as a *functional analysis*. In this step, each of the component parts listed in step 3 is identified as to the functions it serves. Assuming that a proper and thorough set of functions is listed in step 1, the functions that each part serves should be traceable to one of the functions listed in that step. Step 5 of the value engineering process is illustrated in Figure 7.5.

Value Engineering: Example Solution (Interior Door)		
Step 5. Identify Function of Each Part		
Part	Functions	%
A. Door—hollow metal	1. Provides access	20
	2. Insulate sound	20
	3. Facilitate safety	10
	4. Provide aesthetics	10
	6. Regulate temperature	30
	7. Provide durability	10
B. Handles	1. Provide access	10
	3. Facilitate safety	10
	5. Provide security	80
C. Door stops	3. Facilitate safety	100
D. Mountings—hinges	1. Provide access	40
	3. Facilitate safety	60
E. Lock	1. Provide access	20
	5. Provide security	80
F. Closure	1. Provide access	50
	5. Provide security	50
G. Finish on door	2. Insulate sound	10
	4. Provide aesthetics	70
	7. Provide durability	20

Figure 7:5 Value engineering—Step Five

The purpose of step 5 is to justify the existence of each part or component of the entity being analyzed. If a part listed in step 3 cannot be traced to one of the functions listed in step 1, then one of the following conclusions can be reached.

1. The part or component serves no function regarding the intended purpose of the overall designed entity
2. The value engineering team omitted the identification of a function in step 1

The first conclusion leads to an obvious design modification, the deletion of the part that cannot be related or traced to a function. Often the implementation of the value engineering process will lead to the conclusion to delete some cost elements or parts. Unnecessary finish material; excess hardware, such as glass panes in doors; and unused or unjustified floor space or partitions could be examples of items that cannot be justified.

The second conclusion means that the part or component in question has marginal or no value to the overall design. The fact that the value engineering team overlooks a function is usually a good indication that the function is minor and has little or no value to the intended purpose. Although this is usually the case, it is possible that the value engineering team simply was negligent in determining functions. If this is the case, the team needs to revise step 1 before proceeding.

More likely than listing parts that cannot be traced to functions is that several high-cost parts will be traceable to functions that were equated with low value in step 2 or vice versa. In other words, there may be a poor matching of cost and value. If this is the case, the performance of step 5 along with steps 6 and 7 will make this apparent to the value engineering team.

It should be noted that in doing step 5, several of the parts or components listed in step 3 will serve more than one of the functions listed in the first step. For example, in the exercise shown in Figures 7.1–7.7, the handles listed in step 3 are to serve the access, safety, and security functions. The door itself is traced to six different functions.

When a single part serves more than one function, the cost of the part must be allocated, on a percentage basis, to the functions it serves. In other words, how much of the cost or value of a part is allocable to each function? Although this may seem open to much subjectivity, it is intended that the analysis of the value engineering team result in an objective allocation process.

As noted in an earlier paragraph, step 5 is the hardest and most subjective step. The value engineering team will struggle in its attempt to identify purpose or function of some parts or components. However, it is also true that if the value engineering team is struggling with the step, it is likely that the most learning is taking place. If it is hard to do the step it is probably because the step needs to be done. It would be easy to omit the identification

The Value Engineering Process—The Seven Steps 161

of parts to functions. However, without doing this, one can raise the issue if an optimal design, one that matches costs to owner's identified needs, can be accomplished. Step 5 is the "nuts and bolts" of the value engineering process.

Step 6—Calculate Cost per Function

Step 6 is a purely mathematical, mechanical step. The step can be performed by hand calculations or by a calculator or computer. There is no brainstorming or creativity required in this step. The results of steps 4 and 5 are combined in step 6. This step is illustrated in Figure 7.6.

Value Engineering: Example Solution (Interior Door)

Step 6. Calculate Cost per Function

1. Provide access	A.	Door—hollow metal	$18.00
	B.	Handles	4.60
	D.	Mountings—hinges	10.40
	E.	Lock	3.60
	F.	Closure	30.50
			$67.10
2. Insulate sound	A.	Door—hollow metal	$18.00
	B.	Finish on door	.80
			$18.80
3. Facilitate safety	A.	Door—hollow metal	$ 9.00
	B.	Handles	4.60
	C.	Door steps	17.00
	D.	Mountings—hinges	15.60
			$46.20
4. Provide aesthetics	A.	Door—hollow metal	$ 9.00
	B.	Finish on door	5.60
			$14.60
5. Provide security	B.	Handles	$36.80
	E.	Lock	14.40
	F.	Closure	30.50
			$81.70
6. Regulate temperature	A.	Door—hollow metal	$27.00
			$27.00
7. Provide durability	A.	Door—hollow metal	$ 9.00
	G.	Finish on door	1.60
			$10.60

Figure 7:6 Value engineering—Step Six

In step 5 each component of the overall design entity being analyzed was traced to one or more of the functions listed in step 1. In step 4 the total cost of each part was determined as a function of its material and labor or fabrication cost. In step 6 these individual part costs are distributed to the function or functions that the part serves as identified in step 5. For the door handle cost of $46.00 shown in Figure 7.4, $4.60 is allocated to the access function, $4.60 to the safety function, and so on, as shown in the figure. This cost distribution is made based on the fact that in step 5 the value engineering team allocated 10 percent of the handle part to the access function, 10 percent to the safety function, and 80 percent to the security function.

Once the costs of each part are distributed to individual functions, it is just a matter of bookkeeping to gather all of the part costs traceable to a given function and calculate the *cost per function*. For example, the cost of the access function is $67.10. This cost is composed of five separate part or component costs.

Step 6 determines the cost per function of the proposed design. This functional cost estimate provides the base for the performance of the seventh and final step in the value engineering process.

Step 7—Evaluate and Modify the Proposed Design

Step 7 is the brainstorming or creative step. It is not a step that can be easily taught. The objective of step 7 is to critique the cost of the functions performed by the entity being analyzed, with the objective of improving the design. To do this, the value engineering team may simply ask questions like "Is this too much cost for this function?" "Is there another way of satisfying this function with an alternative design or use of materials?" It is intended that the setting out of a functional cost estimate will promote these types of questions with the possibility of recognizing and appraising alternatives. Obviously the success of this brainstorming step is dependent on the creativity and experience of the value engineering team.

To help promote brainstorming, it is useful to compare either the costs or percentages of costs of each of the functions as determined in step 6 with the cost or percentage of costs per function as perceived by the value engineering team in step 2. In step 2 the value engineering team (in conjunction with input from the potential project owner) made an initial estimate of the relative worth of each function. These estimates can be multiplied by the total cost as determined in step 4 to yield a perceived function cost or value. Often this initial estimate is a logical and defendable distribution of the relative value of each of the functions to the potential project owner.

By comparing what the value engineering team thought each function was worth to the potential project owner to the actual cost per function as determined in the value engineering process, attention may be drawn to poor matches of cost versus value. This comparison is illustrated in Figure 7.7.

Implementing Value Engineering

Value Engineering: Example Solution (Interior Door)				
Step 7. Evaluate and Modify the Proposed Design				
Function	Initial $	Estimate %	Actual per $	VE %
1. Provide access	119.70	45	67.10	25.2
2. Insulate sound	53.20	20	18.80	7.1
3. Facilitate connection	26.60	10	46.20	17.4
4. Provide aesthetics	26.60	10	14.60	5.5
5. Provide security	13.30	5	81.70	30.7
6. Regulate temperature	13.30	5	27.00	10.1
7. Provide durability	13.30	5	10.60	4.0
	$266.00		$266.00	100.0

Figure 7:7 Value engineering—Step Seven

Attention is drawn to the fact that the security function was given a 5 percent value in step 2. However, actual calculation indicates that approximately 30 percent of the estimated cost of the entire door is traceable to this function. This may draw the attention of the value engineering team to the use of alternative door-lock hardware or perhaps even deletion of the lock mechanism altogether.

The fact that the anticipated or perceived worth or value of a function does not match the actual functional cost or cost percentage does not mean that the design is inadequate. It could mean that step 2 was performed incorrectly. Similarly, just because the actual cost of a function as set out in step 6 is less than the anticipated costs as determined by multiplying the percentages in step 2 times the total budgeted cost in step 4 does not mean that the design cannot be improved. However, obvious discrepancies between value and cost deserve further analysis. This analysis may or may not lead to an improved design. Without an analysis of alternatives, the ability to prepare an optimal design in regard to project cost, time, and quality is purely coincidental or too dependent on the skill of the individual designer. The value engineering process is justified based on the assumption that the more alternatives that are considered the more likely the potential for an improved design.

IMPLEMENTING VALUE ENGINEERING

Knowledge of the seven steps of value engineering will not in itself ensure the optimal results from the use of the process. Identification of items to be analyzed for which value engineering will yield significant cost savings, identification of individuals to be part of the value engineering team, and rules or policies to be followed by the team in carrying out the value engineering all relate to the success or failure of the process. The CM firm should have the responsibility for addressing the above issues as well as being a member of the value engineering team. Some of the issues relevant to a successful value engineering process are discussed in following paragraphs.

The Project Owner's Role

All too often a project owner is critical of the project after it is completed. This is evidenced by statements from project owners like "If only I could have done it over, I would have made this room bigger," or "I would have preferred this type of finish material," and so on.

Part of the reason is that either the project owner or project designer never bothered to place enough emphasis on needs before or while the project was in the design phase. Instead, the potential project owner let the design team determine the project owner's needs.

The omission of a thorough analysis of a potential project owner's needs could be prevented if the project owner sets them on paper before or during the design phase of the project. This can be effectively performed by having the project owner be part of a value engineering process. The project designer and/or CM firm can request the project owner to participate in steps 1 and 2 of the value engineering process.

Specifically, forcing the project owner to identify the functions to be served by the proposed buildings or parts of the building and also to identify the relative value of each function can in great part dictate the project design elements. Equally important, it is possible that without performing this step, the potential project owner has not thought about the functions or relative worth. For example, the project owner may have a preference for certain finishes. However, once the cost of the finishes relative to their worth to the overall project is realized, the owner may decide against them.

The layout of the space requirements for a proposed project and the quality of construction and budget to be allocated for various items is dependent on the potential project owner's identification of functions to be performed and the relative worth of the functions to the owner. For example, the potential owner of a proposed hospital project has to set out the importance of the various functions performed in the hospital as part of the project design process. Questions such as how much emphasis to place on out-patient care, special high-technology services such as open-heart surgery, and laboratory capabilities all dictate different project drawings and contract documents. Although it is ideal to emphasize all possible services, the owner of the hospital complex, given budget constraints, will likely have to seek trade-offs and set out priorities. These trade-offs and priorities become the basis for the preparation of contract documents.

Constructing the Value Engineering Team

Much of the argument for the implementation of value engineering centers around the concept that "two heads are better than one." It follows that three heads are better than two, and four heads are better than three. This concept is only valid if the two heads have different experiences and complementary skills.

It follows that an effective value engineering process results when more than one individual is involved. In the simplest process, the value engineering team may consist of the proposed project's construction manager and the project's designer. The construction and management skills of the construction manager are intended to complement the design skills of the architect/engineer.

As a project grows in scope and size, an expanded value engineering team is justified. Value engineering teams that include a half-dozen individuals may comprise the team for a several million dollar project. Ideally, the expanded value engineering team members have diverse experience and skills and view the proposed project from different perspectives. For example, for a building construction project, the value engineering team might be comprised of the following:

Construction manager	Material supplier
Designer	Potential user of the project
Owner's representative	Custodian
Contractor	

Each of the above individuals can add valuable input to the project team. For example, the contractor might inject information regarding costs of installing proposed materials, the material supplier can input knowledge of the cost and availability of proposed materials, a potential user of the project may define spacial and functional requirements, and the custodian may add valuable input regarding life-cycle costs.

The roles of the project designer and the potential project owner in the value engineering team have been previously discussed. The construction manager serves as the captain of the team. In addition to adding knowledge of alternative methods and materials, the construction manager draws out the input of the other team members, facilitates the coordination and cooperation of the individual members, and creates an environment in which each of the team members can be creative and appreciative of the views of one another. In addition, the construction manager's setting out and enforcing of certain guidelines or rules can enhance creativity and cooperation. These procedures or rules are discussed in the following section.

Ground Rules for Implementing VE

The purpose of value engineering is to draw out the creativity of the users with the objective of increasing the value of the evolving design. The amount of creativity and related benefits derived by the use of value engineering are partly dependent on the creativity of the participants. However, the ability to evaluate and improve a proposed project design creatively is also dependent on the manner in which the value engineering process is performed. Certain do's and don'ts should be imposed by the CM firm that is leading the value engineering team. Included in these do's and don'ts are the following:

> The CM firm should force the team to make decisions and to serve as a mediator regarding the opinions of the individual value engineering team members.
>
> The CM firm should promote the involvement of each value engineering team member. It's role is to draw out the views of the participants.
>
> The CM firm should determine what items or aspects of the contract drawings are to be evaluated and should budget and control the time the value engineering team spends on each item.
>
> The CM firm should be responsible for the value engineering follow-up, that is, the implementation of the findings or recommendations of the value engineering team.
>
> The CM firm should promote creativity through details such as the use of action verbs in step 1 of the value engineering process.
>
> Creativity is promoted by means of brainstorming and is not to be constrained by any team member. For example, one rule to be followed is that no team member is allowed to ridicule a suggestion made by another team member.

The justification and benefit from each of the first four value engineering ground rules is obvious. The CM firm should be in charge of the process. It is to exhibit leadership and direction to the entire value engineering team.

The last two ground rules deserve more explanation. The concept of promoting the use of action verbs in step 1 of the value engineering process relates to the fact that the use of grammar and language has an impact on an individual's creativity. Action verbs like *directs*, *holds*, *carries*, and so on, provide more creativity than the use of non-action verbs like *provide* and *establish*. Although a detail such as the use of some verbs in defining functions versus others may seem too picky, the student of the use of language would differ. Some words result in an individual reacting, other words tend to result in no reaction and thus no creativity.

Value engineering can be categorized as a brainstorming exercise. Often what appears to be an absurd suggestion is in fact an excellent recommendation that results in a more optimal project design. Simple suggestions like "Why use a door, why not just leave an open entry way?" or "Why not tear out the wall and provide moveable partition walls when needed?" may seem absurd, but in fact may make good sense.

Not all brainstorming suggestions are feasible. Occasionally, a value engineering team member will make an unfeasible suggestion. If this individual is ridiculed for the suggestion, he or she may never again make another suggestion, one that could have merit. The fear of being ridiculed for one's thoughts or recommendations is a real constraint to creativity. It follows that a simple but important ground rule of value engineering is that no member be allowed to ridicule another member for his or her input to the process. Some of the best design alternatives are those that at first seem unacceptable.

EXERCISE 7.1

DIRECTING THE VALUE ENGINEERING EFFORT

One of the more difficult aspects of applying value engineering during the design phase is the determination of what items or systems should be analyzed. The construction manager and the value engineering team cannot afford to value engineer every building element like light fixtures, and doors. Assume a project's design team currently has detailed construction drawings prepared for a proposed 7-floor, $6 million, 120,000 square foot office building. Because of the project owner's limited budget, the construction manager and design team are asked to value engineer the drawings with the objective of reducing $500,000 from the estimated cost of the project. Assuming the construction manager is to work with the designer on the value engineering team, set up steps to be used in determining what items or building systems should be focused on to reduce the cost of the project by $500,000.

EXERCISE 7.2

SEVEN STEPS OF VALUE ENGINEERING

Assume that several different interior wall systems are being proposed by a project design team for an office building. Assume that one of these systems is a wall system that you are familiar with. Perform a value engineering analysis of an interior wall system for an office building using that wall as the proposed wall. Perform each of the seven value engineering steps.

8

Project Planning and Scheduling

THE NEED FOR PLANNING AND SCHEDULING

Project planning and scheduling are the most basic CM services. The reason for many construction projects not adhering to a forecasted time or cost budget relates to the lack of an existing formalized project plan or schedule or the lack of implementation and monitoring of the plan or schedule.

Formalized project planning and scheduling imply the use of various management techniques, including bar charts, critical path method techniques, or other more sophisticated methods. Some would argue that the manager of the construction process can properly plan, schedule, and monitor the plan and schedule without a formalized plan or schedule. However, given the complexity and number of resources relevant to completing most construction projects, this is questionable. It can be easily argued that a picture is worth a thousand words.

The preparation and implementation of a formalized, on-paper project plan and schedule can provide the CM several benefits, including analysis of the optimal sequencing of phases of the project, providing a base communication document that can be used as a means of integrating all entities, such as the project design and construction teams, and serve as a means of measuring progress. Other benefits of the prepared project plan and schedule are that they can be used to pinpoint each entity's responsibilities and to set out goals for the CM firm as well as other project entities.

The CM firm's involvement in the preparation and implementation of a formalized plan and schedule is not limited to applications to the construction phase of a project. The use of a project plan and schedule is equally relevant to the preconstruction phase of a project. Setting out and enforcing the completion of various design phases of a project and material purchase order dates are fundamental to satisfying overall project time and cost constraints. The preparation and implementation of a formalized, on-paper project plan and schedule are also crucial to the CM firm's ability to phase or fast-track and package construction contracts effectively. The relationship of project phasing and packaging to project planning and scheduling is discussed in the following chapter.

The CM firm by itself cannot implement formalized project planning and scheduling for the successful time and cost completion of a project. The CM firm must get the involvement and commitment of all of the project entities if its efforts are to be fully beneficial. The CM firm provides the leadership in the process. However, the CM firm must strive to have each of the entities contribute and adhere to the project planning and scheduling functions.

Numerous formalized project planning and scheduling techniques are available to the CM firm. Both bar charts and the critical path method (CPM) are two techniques that are the easiest to use and least sophisticated of the many available. The fact that they are simple does not imply that they are not compatible with the CM firm's or construction industry's needs. Simplicity enhances multiple uses of a technique, a fundamental requirement of planning and scheduling techniques for the construction industry. It should also be noted that any technique or model is only as good as the accuracy of the data. Erroneous data input will result in erroneous output. Perhaps of all the project planning and scheduling techniques available to the CM firm, the bar chart model and the critical path method require data that are most compatible with what is normally known by the CM firm or other project entities.

The simplicity and the availability of the data required by the bar chart model or CPM result in both of these models being most compatible with the CM firm's planning and scheduling objectives. The mechanics of these techniques, with emphasis on CPM, along with several scheduling rules or procedures, are presented in this chapter. Application of bar charts and CPM relevant to project phasing or packaging are discussed in the following chapter.

THE BAR CHART MODEL

Until the early 1950's, the construction project planner was limited to the use of the bar charts in modeling projects. Owing to the simplicity associated with developing a bar chart, they are still often used in project planning.

The bar chart provides a method of integrating the project phases into an overall plan. In addition to showing an estimated start and finish time for each phase, the bar chart provides for calculating the time at which various resources are needed for the project plan and provides the manager with a means of recording the actual progress of the project. By recording the actual progress, it is possible to quickly determine whether the construction project is progressing according to plan.

A typical bar chart is shown in Figure 8.1. The phases of the project are listed vertically. In addition to columns for the quantities of work to be

The Bar Chart Model

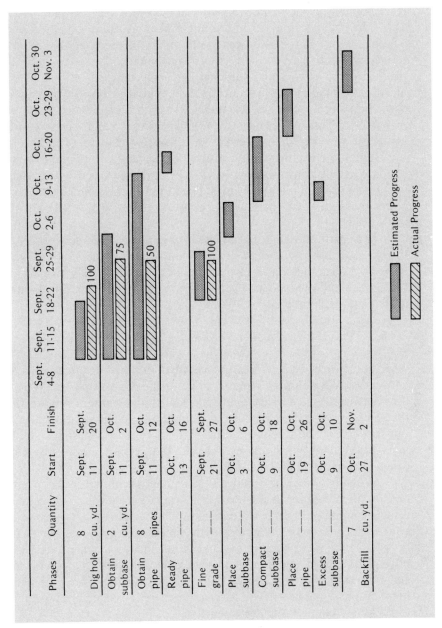

Figure 8:1 Example of a bar chart

performed for each phase and for the start and finish time for each of the phases as determined by the project plan, a time scale is provided by columns corresponding to the months of the year.

The planned time for performing a phase is shown by a horizontal bar. The bar begins at the estimated starting date of the phase and ends at the estimated finish date. In addition to the horizontal bar that indicates the planned phase there is an additional bar (cross-hatched) that is used for indicating the progress of the project. At the end of each of these cross-hatched bars, the percentage of work completed is indicated. If phases exceed their planned duration, it is beneficial to replan the project in progress since individual overruns may influence the performance of following phases.

The bar chart helps the construction manager to plan and control the building of the construction project. Owing to its visual effectiveness, and because of the simplicity of its preparation, several construction managers prefer to use the bar chart over a network model such as the critical path method.

Bar charts, although widely used, have many limitations. In particular, the bar chart does not clearly show the interdependencies among the project phases. As a result, the course of action the construction manager should take is not indicated when a particular phase's actual duration is not equal to its planned duration.

NETWORK MODELS AND PROJECT PLANNING

The use of networks has greatly facilitated the task of project planning. A network is a diagrammatic representation of the project's phases. Whereas the bar chart is restricted in its ability to identify the complex interrelationships that exist among project phases and to indicate optimal project phase timing, calculations may be made by means of the network model that indicate optimal phase timing decisions. As a result, the network model has become a widely used project planning tool.

When constructing a network to represent a project, either arrow notation or circle notation is used. An arrow notation network is constructed by using arrows to represent project phases. The relation among phases is represented by connecting the arrows with circles (often referred to as *network nodes*). The nodes represent events that are points in time. For example, a node placed on the end of a phase represents the event "the phase is finished and following phases may start." Figure 8.2 illustrates several small arrow notation phase networks. The letter on each of the phase arrows corresponds to the phase name or identification.

The use of the arrow notation network for modeling a project sometimes necessitates the use of a dummy arrow to properly model the project phase logic. Consider the following logic:

Network Models and Project Planning

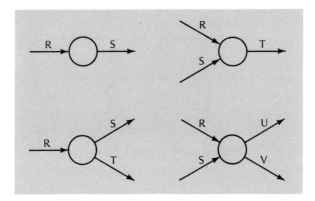

Figure 8:2 Arrow notation CPM

Phase C follows Phase A
Phase D follows Phase B
Phase C follows Phase B

When using the arrow notation network, the only correct way to represent this phase logic is as shown in Figure 8.3. The dashed arrow in the figure is referred to as a dummy arrow. A dummy arrow has zero cost and zero time associated with it. It is used to represent the proper logic that exists among project phases.

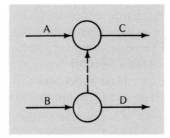

Figure 8:3 Use of the dummy activity

The dummy arrow is also used to avoid two phases that start at the same node (event) and terminate at yet another node (event). To avoid the loop formed by the two phases a dummy arrow is introduced. The use of the dummy arrow for this purpose is shown in Figure 8.4.

A circle notation can also be constructed to represent the logic among a project's phases. In the circle notation network, phases are represented by circles. The logic among the phase circles is shown by connecting the circles

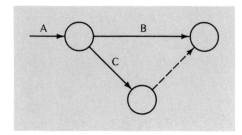

Figure 8:4 The dashed arrow to avoid a loop

with arrows. As in the arrow notation network, the actual length of the arrow is of no significance. Figure 8.5 illustrates some project phase relationships by means of a circle notation network. The letter in each circle corresponds to the phase name or identification. When using a circle notation network to model project phase logic, there is no need to introduce a dummy arrow. However, to define a single starting datum and ending datum, it is sometimes convenient to create Start and End activities as shown in Figure 8.5.

Obviously, a given construction project can be modeled by either an arrow notation or a circle notation network. Arrow notation has the advantage of providing a better visual representation of the project phase logic than the circle notation network. When project phase logic becomes complex, it becomes difficult to follow a circle notation network visually. Another advantage of the arrow notation network is that it is relatively easy to transform the network into a time scale. This will be discussed in a later section. Owing to its visual effectiveness, the arrow notation network will be used in this book to represent project phase logic.

The circle notation network has the advantage of being easier to construct than the arrow notation network. There is no need for the use of the dummy activity. The circle notation network is especially advantageous when changes in the logic must be made. These logic changes are frequent when performing resource allocation.

In project planning, time or phase duration is the most important factor. These phase durations are assumed to be determined in the project phase planning task.

THE CRITICAL PATH METHOD (CPM)

One of the first network models developed, and presently the most widely used construction project planning model, is the critical path method (CPM). The model was originated by James E. Kelly of the Remington Rand Corporation and Morgan Walker of the DuPont Company. Their original

The Critical Path Method (CPM)

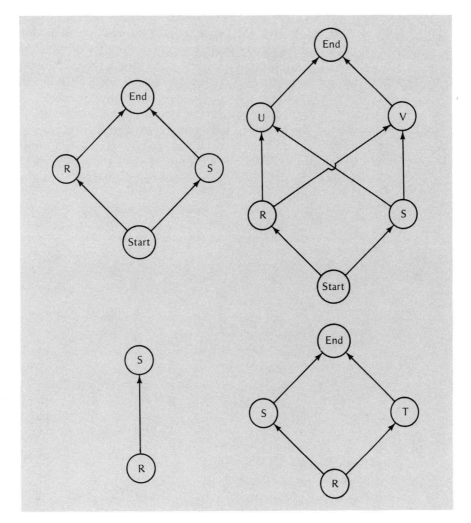

Figure 8:5 Circle notation CPM

intent was to develop a planning and scheduling tool for the construction and maintenance of chemical processing plants.

CPM has become so widely accepted as a planning and scheduling tool that many construction project owners and construction managers require contractors to submit a CPM plan of the project along with their bid. Although CPM may be used to model any type or size of project, the benefits received from its use are best realized on projects that are rather large in terms of complexity and the number of required project phases, and projects that are "one of a kind." For very small projects that have been constructed several times in the past, so that the manager knows a best way of performing the project's phases, it may be unfeasible to spend the time or money associated with developing a CPM model. However, in general, the CPM model can be constructed for a project so that the benefits the construction manager receives from the model's use far outweigh its cost of preparation.

Besides some of the more obvious benefits from the use of CPM which are discussed in the following section, a CPM approach to project planning is useful in that it forces the construction manager to divide a proposed project into several phases. This in itself is useful, since the construction manager becomes better acquainted with the various components of the project.

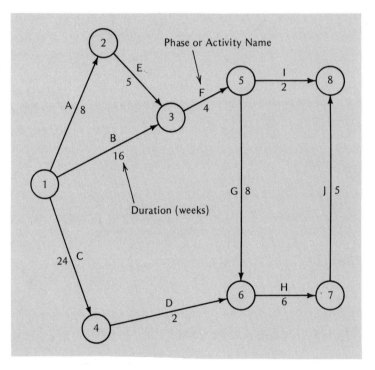

Figure 8:6 Example of a CPM diagram

The CPM model assumes a determined project phase network structure and phase parameter (duration). Given a phase network, such as that shown in Figure 8.6, CPM assumes that if activities B and E are completed, activity F can and will occur. There is no chance (a probability of 0) that activity F will not occur, given that activities B and E are completed.

The variable in question is duration. Given activity durations as determined in the project phase planning, one of the objectives of CPM is to determine project duration. When using CPM, it is assumed that each project phase's duration is known with certainty. Given that a phase has a duration of 10 days, there is no probability of the phase taking 9 or 11 days. This is the assumption of the CPM calculation procedure.

When the project network is drawn in arrow notation, it is common to place duration of a phase alongside the phase arrow. The units of time are often not placed on the arrow unless the time units vary from one activity to the next. Rather, the time unit used in the network may be stated in the upper-right-hand corner of the network's drawing.

Given the deterministic nature of the CPM network and the phase durations, the project duration is found by merely summing phase durations. However, when adding phase durations, the fact that a phase cannot start until the logic of the network is satisfied must be considered.

CPM CALCULATIONS

Determining the project duration is not the only purpose of the CPM model. Determining a phase's possible delay time and which phases are critical (in that they dictate the project duration) are equally important objectives. To satisfy such objectives, it becomes necessary to generalize the CPM calculation procedure.

Regardless of the size of the project network in question, or the type of network notation, whether arrow or circle, the objectives of the basic CPM calculations and the actual calculation procedure remain unchanged.

There are three objectives to be satisfied by basic CPM calculations. These are as follows:

1. Determine the completion time or duration of the construction project
2. Determine which of the phases of the project are critical, that is, which phases determine the completion time of the project
3. Determine how much time each of the project phases may be delayed without affecting the completion time of the project as determined in the first objective

To satisfy these three objectives, the construction manager must perform the following.

1. Define project phases
2. Determine project phase durations
3. Determine technological logic among phases
4. Perform CPM calculations

Determining project phases and their durations is part of estimating. Determining the technological logic is the responsibility of the construction manager. Let us now consider a simple example project to demonstrate the CPM calculations procedure.

A construction manager must prepare a project plan for a small building. The project is divided into the following phases—preconstruction and construction.

Phase	Phase Indentification
Prepare site drawing layouts	A
Prepare building design—foundation and frame	B
Prepare specialty trade drawings	C
Prepare drawings for finishes	D
Obtain required zoning and permits	E
Construct foundation	F
Construct building frame	G
Perform specialty trade work	H
Finish site work	I
Perform finishes work	J

Observe that each of the phases is assigned a letter. This is done only to shorten the identification of the phases for the CPM network model.

The construction manager next identifies the technological logic that exists among the phases and, recognizing resources and estimates for the project, establishes phase durations. The defined logic and the phase durations are as follows:

> Phase A can be done initially and takes 8 working weeks.
> Phase B can be done initially and takes 16 working weeks.
> Phase C can be done initially and takes 24 working weeks.
> Phase D can be done after C and takes 2 working weeks.
> Phase E can be done after A and takes 5 working weeks.
> Phase F can be done after B and E and takes 4 working weeks.
> Phase G can be done after F and takes 8 working weeks.
> Phase H can be done after D and G and takes 6 working weeks.
> Phase I can be done after F and takes 2 working weeks.
> Phase J can be done after H and takes 5 working weeks.

The phases' durations are all in terms of work weeks. Thus, all calculations and solutions to the three CPM objectives will be in terms of work weeks. Of course, in other projects, phase durations can be given in days, months, and so forth. The important requirement is that a common unit of time be used for all phase durations. The CPM arrow notation network model for the defined construction project is shown in Figure 8.6. The numbers in

CPM Calculations

the nodes are used only to identify the particular events. They have no bearing on the CPM calculations.

The construction manager wants to determine the completion time (in work weeks) of the project, the critical phases, and the possible delay time of the noncritical phases. To satisfy these objectives, CPM calculations must be performed. These calculations are performed in two steps; the first is known as the *forward pass*, and the second the *backward pass*.

Let us consider the forward pass. This is also referred to as the *earliest start time schedule*, since the calculations performed will yield information pertaining to the earliest time that the project phases can possibly start and finish.

The earliest start time calculations may be made in several ways. The earliest event time (EET) of a node is defined as the earliest time that an event can possibly occur. To identify the value of an EET on the project network, its value is shown in a five-sided figure pointing to the right placed next to the event's node, as shown in Figure 8.7. The direction of the EET figure indicates that its value was determined by a left-to-right sweep of the network. From the definition of an EET, the EET of event 1 in the project network can be determined. It is obvious that its value is 0. The project can start as soon as possible.

The *earliest start time* (EST) of a phase is defined as the earliest possible time the phase can start. For example, phases A, B, and C are all equal to 0, which is the value of the EET of the event or node at which they originate. This makes sense in that these activities cannot start until the event preceding them has occurred. The reader should observe that the EST, in addition to any other times discussed, is referenced to the end of the previous week rather than at the beginning of the present week.

The *earliest finish time* (EFT) of a phase is defined as the phase's earliest start time plus its estimated duration. Therefore, the EFT of phase A is 0 plus its duration of 8 weeks. Therefore A's EFT is 8 weeks. The EFTs of phases B and C may be found in the same manner.

To calculate the remaining EETs for the nodes or events in the network, the definition of an EET is used. The definition states that the EET of an event or node is the earliest time the event could occur. An event cannot occur until the phases immediately preceding it have finished. Therefore, the EET of the remaining nodes can be found by those preceding them. For example, node 2 only has one phase immediately preceding it. Therefore, its EET is 8, which is the EFT of phase A. Node 3 has both phases B and E immediately preceding it. The EFT of E is $8 + 5$, or 13, the EFT of B is $0 + 16$ or 16. Therefore, the EET of node 3 is 16. The EETs for the remaining nodes, and the ESTs and EFTs for the remaining phases are calculated in the same manner as described for the previous event and phase times. The EETs for all of the events are shown in Figure 8.7. Phase ESTs and EFTs are shown in Figure 8.8.

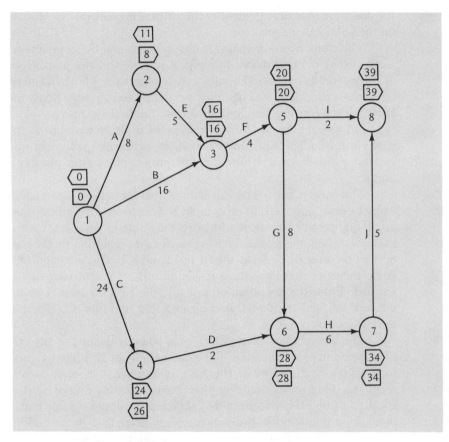

Figure 8:7 EETs and LETs

1 Activity	2 DUR	3 EST	4 EFT	5 LST	6 LFT	7 FF	8 FFP	9 TF
A	8	0	8	3	11	0	3	3
B	16	0	16	0	16	0	0	0
C	24	0	24	2	26	0	2	2
D	2	24	26	26	28	2	0	2
E	5	8	13	11	16	3	0	3
F	4	16	20	16	20	0	0	0
G	8	20	28	20	28	0	0	0
H	6	28	34	28	34	0	0	0
I	2	20	22	37	39	17	17	17
J	5	34	39	34	39	0	0	0

Figure 8:8 Solutions to CPM calculations

CPM Calculations

The EET of node 8 has special significance. This event time, which corresponds to the end of the thirty-ninth week, is equal to the completion time of the project. By definition of an EET, this is the earliest possible time for the event to occur. Since it is the last node of the network, it represents the end of the project. Therefore, one of the objectives of the CPM calculations has been satisfied, that is, the completion time of the project is found to be 39 working weeks.

Additional information can be obtained from the completed earliest start time schedule. The EET of each event or node is determined by the phase or phases that have the largest EFT. As such, not all the phases determine the value of the EET of the node at which they terminate. The phases that do not can be delayed without affecting the EET. The amount of time a particular phase may be delayed without affecting the EET of the node at which it terminates is the difference between the EET at which the phase terminates and the phase's EFT. This possible delay time is often referred to as *free float* (FF). The free float of phase A is 8 — 8 or 0 weeks. Phase A's free float is zero because the phase's EFT determines the EET for node 2. The free floats for all of the phases are shown in Figure 8.8. Although giving a measure of possible phase delay time, free float is often a conservative measure of a phase's actual possible delay time. Therefore, the objective of determining possible phase delay times is not completely satisfied by the earliest start time calculations.

In addition, the determination of the critical phases (the phases that dictate the project duration), remains to be satisfied. Admittedly, an inspection of the network may identify the critical phases. However, such a visual identification becomes difficult as the number of phases in the network increases. It is difficult to determine the critical phases in the network. It is desired to determine the critical phases by means of calculation, rather than trying to determine them by a visual inspection of the network.

Note that if one desires to ignore the existence of events on the arrow CPM network, one may still perform the required CPM calculations necessary to satisfy the three objectives. The EST of a phase merely becomes the maximum EFT of the phases that immediately precede it. The ESTs of the originating phases of the project are equal to 0 by definition. The earliest finish time (EFT) of an activity is merely the sum of its EST and the phase's duration. The project duration is equal to the maximum EFT of the phases that terminate the project. Finally, the free float of a phase is the difference between the phase's EFT and the EST of the phase immediately following it.

Thus, it becomes possible to ignore the existence of event times (EETs) when obtaining the project phases' ESTs, EFTs, and free floats, in addition to the project duration. Depending on whether or not one discusses events as being part of the CPM network, one may or may not refer to event times (EETs) as being part of the CPM calculation procedure.

The backward pass on the CPM network is referred to as the *latest start time schedule*. This is because the calculations performed yield information on the latest times phases can start and finish. As in the earliest start time calculations, the latest start time calculations are made by using several definitions.

The *latest event time* (LET) of an event is defined as the latest possible time the event can possibly occur without delaying the completion time of the project (as determined either by the earliest start time schedule or as fixed by the planner). From this definition, the LET of node 8 in the network in Figure 8.7 is determined as 39, by the EET of node 8. This means that the latest possible time for event 8 is the latest time event 8 can occur to complete the job in 39 weeks. The datum, or starting node, for the backward pass is the terminating node for the forward pass. The LETs for nodes are placed in five-sided figures pointing to the left, as shown in Figure 8.7.

The *latest finish time* (LFT) of an activity is defined as the latest possible time that a phase can finish without delaying the completion time of the project. The LFT of a phase is equal to the LET of the node or event at which the phase finishes. This is true because of the definition of LET. As such, the LFTs of phases I and J in Figure 8.7 can be found directly to be equal to the LET of node 8 (that is, 39), which corresponds to the completion time of the job.

The *latest start time* (LST) of a phase is equal to the phase's LFT minus the duration of the phase. In the example in Figure 8.7, the LST of phase I is 39 — 2, or 37, and the LST of phase J is 39 — 5, or 34. As in the earliest start time schedule, these times refer to the end of the given work week.

The LETs for the remaining nodes may be determined from yet another definition of LET. The LET of a node or event is equal to the minimum of the LSTs of the phases that originate at that node. Therefore, the LET for an event is dependent on the phase that originates at the event. For node 7, phase J is the only phase originating at the node. Therefore, the LET of node 7 is equal to the LST of phase J, which is 34. Similarly, the LST for node 6 is 28.

Phases I and G originate at node 5. The LST of I is 39 — 2, or 37, and the LST of G is 28 — 8, or 20. Therefore, the LET of node 5 is 20. The LETs for the remaining nodes, and the LSTs and LFTs for the remaining phases can be found from the stated definitions. The LETs are shown in Figure 8.7 and the LSTs and LFTs in Figure 8.8.

Obviously, every phase cannot determine the LET of the node at which it originates. The phases that do not can be delayed without affecting the value of any LET of the network. The amount of time that a phase can be delayed is the difference between the phase's LST and the LET of the node at which the phase originates. This possible delay time is sometimes referred to as a *free float prime* (FFP) or *backward float*, versus free float or forward float as determined in the forward pass. In our example in Figure

CPM Calculations

8.7, phase I has a free float prime of 37 — 20, or 17 weeks. Phase J has a free float prime of 34 — 34, or 0. Phase J has a free float prime of 0 because it determines the LET of its originating node. One might think of free float prime as the possible amount of time that the start of a particular phase can be delayed; free float is the possible amount of time the finish of a particular phase can be delayed. The free float primes for all the phases of the project are shown in Figure 8.8. Free float primes give a measure of possible phase delay times. However, as in the case of free floats, this measure is often too conservative.

It should also be observed that as in the earliest start time schedule calculations, one can ignore the existence of event or node times associated with the latest start time schedule. The latest finish time (LFT) becomes the minimum LST of the phases that immediately follow it. The LFTs of the terminating project phases are defined as being equal to the project duration as determined by the earliest start time schedule, or as fixed by the project planner. The latest start time (LST) of a phase is defined as its LFT minus its duration. Finally, the free float prime is equal to the maximum LFT of the phases that immediately precede the phase in question, minus the phase LST.

Free float and free float prime, although useful in several resource allocation algorithms, do not indicate the actual possible delay time associated with a phase. This is a function of both the earliest start time schedule and the latest start time schedule.

Another type of project float is referred to as a *total float*. Total float is a function of both the earliest and latest start time schedules. Total float is defined as the difference between the LST and EST of the phase, or the difference between its LFT and EFT. Either of these calculations yield the same total float value. In the example in Figure 8.7, phase I has an LST of 37, whereas its EST is 20. This means that it may start at the end of week 37 and not delay the completion time of the project. On the other hand, phase I may start as early as the end of the twentieth week. As such, phase I may be delayed as much as 17 weeks. This is proved by a calculation of the total float, which yields 37 — 20, or 17 weeks. If the calculation is based on the LFT and EFT of the phase, we find the total float to be 39 — 22, or 17. The total floats of the remaining phases of the network are shown in Figure 8.8. The total float of a phase is, therefore, the amount of time the phase can be delayed without delaying the project completion time.

If a phase has an EST equal to its LST, then it must begin as soon as possible. It cannot be delayed. Of course, this is also true if a phase's EFT is equal to its LFT. The phases that cannot be delayed (so that the projected completion time is not delayed) are called *critical phase activities*. Critical phases that form a continuous path from the start to the end of the project are known as the project's *critical path*. A project always has one or more critical paths. To control the duration of the project, the construction manager must direct his or her efforts to the phases on the critical paths. Admittedly, if phases that have total float are not completed within their allowable float times, they too can become critical. In other words, for a project to be completed according to the schedule determined by the phase technological logic, each and every phase must not exceed its calculated total float.

At this point, the construction manager has satisfied all three objectives. The possible project completion time, the critical phases, and the possible delay time of the phases have been determined. In the case of the building project, the CPM calculations have yielded a possible completion time of 39 weeks. The critical phases are phases B, F, G, H, and J. Phases A, C, D, E, and I have possible delay times equal to their total float (shown in Figure 8.8).

The CPM project network and the CPM calculations are not limited to determining project duration, the critical phases, and the phases' allowable delay times. Actually, the CPM model enables the construction manager to plan and schedule the project phases optimally, subject to resource allocation considerations. The CPM model facilitates the distribution of the construction manager's resources. The optimal time to perform a phase is not only a function of the project phases' technological logic, but also a function of the resources to be used to perform the project.

Finally, as previously mentioned, calculations performed on a circle notation CPM model will yield the same information as calculations performed on an arrow notation model, if the models represent the same project. A CPM circle notation network for the discussed building project is shown in Figure 8.9. Observe that the EETs and LETs are not as meaningful at their locations as in the arrow notation network. This is because the circles or nodes in the circle notation network now represent phases rather than events. As would be expected, the phases' ESTs, LSTs, EFTs, LFTs, and floats are equal to the values calculated by means of the arrow notation network and are shown in Figure 8.8.

CPM is most useful and rewarding for large, complex projects. For these types of jobs the calculations become cumbersome and mistakes are easily made. Therefore, CPM calculations are often performed with a computer. Because of the repetitious nature of the calculations, they are easily programmed. A user merely inputs the phases, their duration, and the logic among the phases. As output, information pertaining to the duration of the project, the critical phases, and the possible delay times is received.

TIME SCALE CPM

To gain a visual representation of the CPM project plan and schedule, it is useful to plot the CPM network on a time scale. Such a time scale network requires the use of arrow notation. In the CPM arrow notation network discussed in the previous section (see Figure 8.7), the length of individual

Time Scale CPM

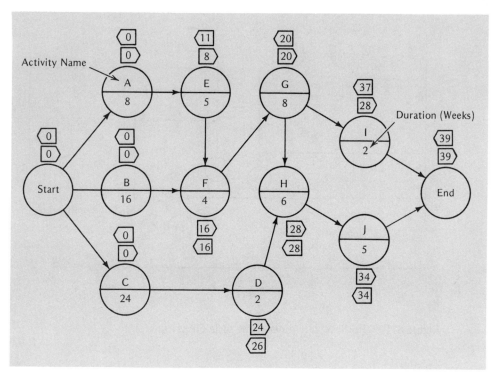

Figure 8:9 Example of a circle notation diagram

arrows was not related to the duration of the phase represented by the arrow. However, when constructing a time scale CPM network, the length of arrows used to represent phases are proportional to the phase durations.

The time scale CPM network is constructed by connecting phase arrows according to their defined logic, and drawing each phase arrow as a straight line and of a length dictated by the phase duration and the time scale (which is placed on the horizontal axis of the time scale CPM). Figure 8.10 shows the previously discussed building project on a time scale CPM network. Note that the network plan and schedule shown correspond to the earliest start time for the project. The dashed arrow following a phase corresponds to the phase free float. The ESTs and EFTs of each of the phases can easily be determined by a visual inspection of the network.

It should be observed that when constructing a time scale CPM, one is actually indicating a project schedule; that is, the exact timing of each of the phases is determined. Since several phases of a project can often be shifted within their allowable float times, one could generate several time scale CPM networks for a single project, each network representing a feasible schedule. For example, Figure 8.11 shows the latest start time CPM time scale net-

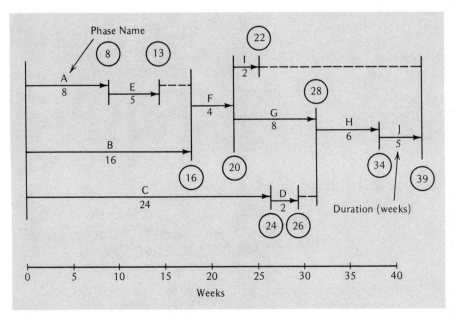

Figure 8:10 Earliest start time—time scale CPM

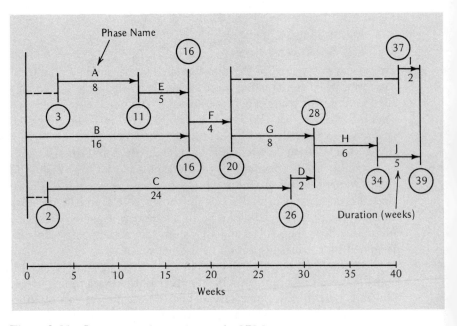

Figure 8:11 Latest start time—time scale CPM

work for the building project. The dashed arrows in the figure represent phase free float primes.

Besides being visually effective, the time scale CPM network proves beneficial as an aid to resource allocation and packaging and phasing work. Owing to the time scale network's visual presentation, it becomes easy to determine resource requirements and conflicts.

EST-LST SCHEDULES AND SCHEDULING RESOURCES

In the previous sections, CPM calculations were performed by means of the earliest start time (EST) schedule, and the latest start time (LST) schedule. It should be recognized that these two project schedules are feasible schedules. Given the duration of each phase, and the resource requirements needed for the duration of each phase, one may easily determine the schedule of required resources for both the EST and LST project schedules.

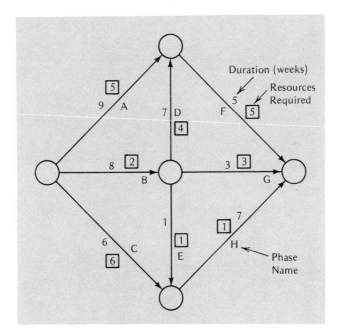

Figure 8:12 CPM with resource requirements

For example, consider the construction project modeled by the arrow notation CPM network shown in Figure 8.12. In addition to the duration of each activity, a resource requirement for each activity is shown. There is only a single type of resource required for the project. The EST time scale CPM

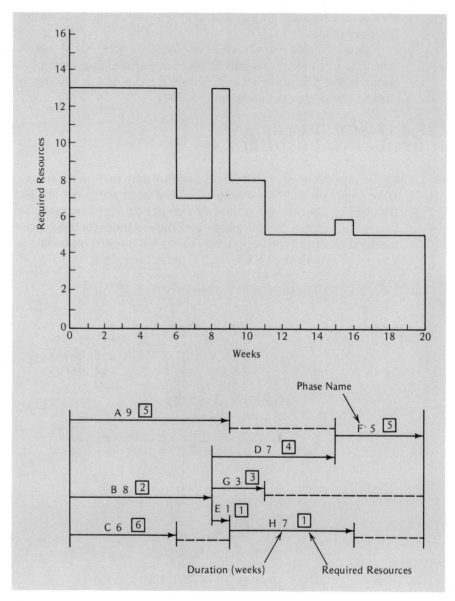

Figure 8:13 Resources and EST schedule

for the described project is shown in Figure 8.13. The project duration is 20 weeks. The dashed arrow, following a phase shown in Figure 8.13, corresponds to the phase's free float. By summing the resources required of the phases that occur on any given week, one may represent the required resources as a

EST-LST Schedules and Scheduling Resources

function of time. The required resources as a function of time for the EST project schedule are also shown in Figure 8.13.

Similarly, an LST network for the project is shown in Figure 8.14. The LST project duration was made equivalent to the EST project duration.

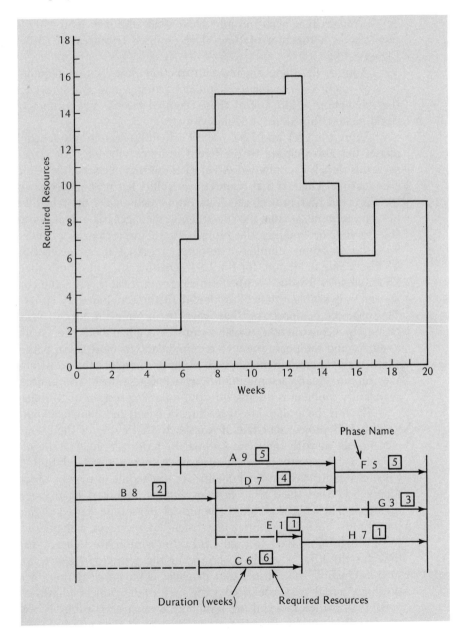

Figure 8:14 Resources and LST schedule

Admittedly, the construction manager may desire a different project duration. If this is the case, it is merely a matter of changing the datum for the calculations. However, the phase logic and phase occurrence remain the same.

The dashed arrow immediately preceding a phase in Figure 8.14, corresponds to the phase's free float prime. By totaling the resources required by the phases that occur on any given week one may represent the project resources as a function of time. The required resources are also shown in Figure 8.14.

Admittedly, there are few construction projects that require only one type of resource. In actuality, a project requires several resources. However, the calculation of any one of these required resources as a function of time could be determined in the same manner.

Thus, the EST and LST project schedules do not only schedule project phases but also indicate two different resource schedules for projects. However, this does not imply that either of these resource schedules is optimal, or even feasible. The fact that resource availability has been ignored in producing the EST and LST project schedules can actually make the schedules unfeasible. For example, assume the construction manager has only 9 resources available to use for building the project. From inspection of Figures 8.13 and 8.14, the maximum number of resources required at any given time for the EST schedule is 13, and, for the LST schedule 16. Thus, both the EST and LST schedules for the described project are unfeasible, since the construction manager's available resources are less than those required by either schedule. This example demonstrates that when performing the task of resource allocation, the construction manager must recognize both the availability of resources and the phase resource requirements. In most cases, both the availabilities and the requirements will be finite; therefore the construction manager cannot usually assume infinite available resources. The limited resource availability problem is discussed in the following section of this chapter.

Even if the availability of resources exceeds the number required by the EST or LST project schedule, it is unlikely that either of the resource schedules associated with them are optimal in terms of project time or cost. Although it is necessary to consider resource costs and project time continuity when evaluating the financial qualities of a schedule, it may be observed from Figure 8.13 that the EST schedule requires an initial large investment of resources. This is because all of the project phases are started at their earliest start times. Such large initial investment of resources will undoubtedly result in high initial cost. In Figure 8.15, the cumulative resource-weeks versus time for the EST schedule is shown for the project. In effect, when using such a schedule, the construction manager is buying float time or risk, since the high initial cost associated with performing project phases as soon as possible and thereby requiring resources as soon as possible, is rewarded by the fact that possible float times are obtained at the end of phases. Thus, if

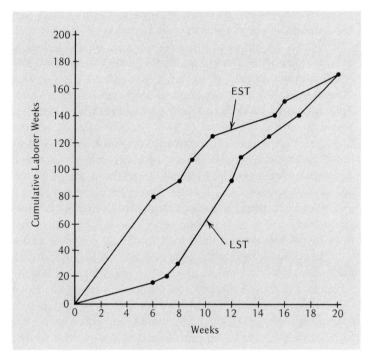

Figure 8:15 Cumulative resources required

a circumstance occurs that delays a phase, and the phase has float, there may not be any increase in project duration.

By using the latest start time resource schedule, the construction manager is delaying high initial resource investments in favor of using phase floats at the beginning of the phases (delaying the start of the phase). The cumulative resource-weeks versus time for the LST schedule of the project described is shown in Figure 8.15. Such a schedule has a lower initial resource investment versus the project's EST schedule, but it has a higher risk because the schedule does not allow for float times at the end of phases. Thus, if a circumstance occurs that results in an increased phase duration, the project duration may have to be extended. Such an increase in project duration often results in added project costs.

In Figure 8.15, it can be observed that both schedules have a project duration of 20 weeks and a cumulative resource-weeks total of 167. It is obvious that this is true. The area bounded by the EST and LST project schedules in Figure 8.15 represents other possible resource schedules for the project. One might view the EST and LST schedules as corresponding to the two extremes, or bounding resource schedules. By shifting phases within their allowable float times, several other project schedules may be produced that do not exceed the project duration of 20 weeks.

It may be observed from Figures 8.13 and 8.14 that both the EST and LST resource schedules are not "level" schedules in that both require a somewhat uneven number of resources throughout the duration of the project. For example, the EST schedule initially requires 13 resources, then drops to 7, then back to 13, then to 8, and so forth. Possibly, a better resource schedule would require a more even, or level, amount of resources throughout the duration of the project. Theoretically, the most level schedule for the project in question would require the use of 8.35 resources per week (167/20). Such a schedule would be a straight line from the origin to the point described as 167 resource-weeks and 20 weeks. Surely such a schedule is impossible. However, it may be possible to approach such a schedule by using 9 resources on some weeks and 8 resources on other weeks. The actual process of averaging the resources required during the project is referred to as *leveling*.

Leveling is facilitated by using float times. By shifting phases within their allowable float times, it may be possible to obtain a desired resource schedule. Leveling is discussed in a following section of this chapter.

Although the EST and LST project schedules may not be the most optimal resource schedules, they provide the construction manager with data for starting the scheduling process and evaluating the benefits of a particular schedule. It should also be noted that, depending on such factors as the cost of the various required resources, their availabilities, and the schedule of project incomes, the EST or LST may actually be most optimal in regard to project time and cost.

PROJECT SCHEDULING WITH LIMITED RESOURCES

A construction manager is often limited in the use of resources when determining a project plan and schedule. Even labor and material, although usually considered as infinitely available can in some instances be limited for a project under consideration. If an additional resource can only be obtained by added cost, it may be advantageous to consider the resource as not being available (limited amount available).

It is common to develop methods for various project phases independent of one another. Thus, even if every project phase is planned within the construction manager's resources, it is more than likely that phases that occur simultaneously will require resources in excess of those available. Therefore, in addition to the technological phase constraints, the construction project planner must consider the limited amount of project resources. Resource

Project Scheduling with Limited Resources

availabilities must be incorporated into the logic of the overall project plan and schedule. If one assumes that phase resources and durations remain fixed, then it is very possible that the phase logic, as determined by technological constraints, may have to be changed to account for the availability of resources.

One procedure that can be used to solve the problem of constructing a project plan subject to limited available resources is to consider, simultaneously, both the technological activity constraints and the limited resources when constructing the project plan. However, this is often very difficult to do for projects that contain several phases, and the phase logic so constructed is often less than optimal.

A more feasible approach is first to construct the network subject only to the technological phase constraints, and then to relieve existing phase resource conflicts (requirements greater than availability) in a fashion that allows for the optimal project plan and schedule. Although the construction manager can have one of many project plan objectives, minimum project duration means that the construction manager wants to minimize uncostly phase delays, maximize productivity, and so on.

One such model or algorithm that attempts to relieve phase resource conflicts while keeping the project duration to a minimum is known as the *resource scheduling method* (RSM). It relieves resource conflicts by focusing on two conflicting phases at a time.

Let us consider the CPM arrow notation network discussed in the previous section and shown in Figure 8.12. The project is shown by a time scale EST CPM model in Figure 8.16. As in the case of CPM shown in Figure 8.12, each phase in Figure 8.16 is assigned a duration and a required number of resources. Let us assume that the construction manager has access to only 9 resources. From an inspection, it may be observed that each phase may be performed, since each requires less than 9 resources, but it may also be observed that the project logic requires more than 9 resources at a point in time. In particular, on day 1 of the project, phases A, B, and C require a total of 13 resources. Thus, there is a resource conflict in that 13 resources are required and only 9 are available. It is necessary to change the existing logic among phases A, B, and C so that the project plan does not require more than 9 resources.

Before addressing the project in question, let us refer to the procedure used in RSM for relieving phase resource conflicts. The RSM model removes a resource conflict that exists between two phases by forcing one of the phases to follow the other. This change of logic is meant to keep the project duration to a minimum. Consider phases X and Y shown in Figure 8.17. Let us assume that the two phases have a resource conflict and cannot be performed simultaneously. The question becomes one of whether phase X

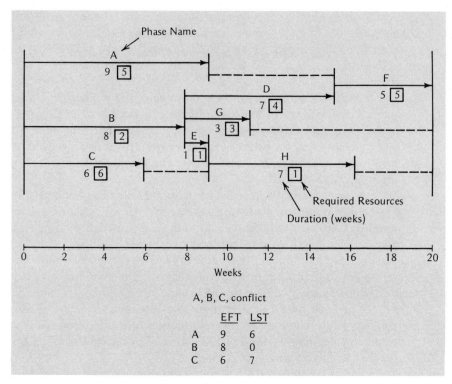

Figure 8:16 Resources limited to nine

should follow phase Y or vice versa. As a result of other phases (not shown), the ESTs, EFTs, LSTs, and LFTs are as follows:

	EST	EFT	LST	LFT
X	9	15	10	16
Y	10	17	14	21

If phase X is forced to follow phase Y, the increase in project duration is 7 weeks, shown as D_{YX} in Figure 8.17. This increase results because the EFT of activity Y is week 17, and the LST (latest time phase can start without increasing project duration) of activity X in 10 weeks. Since phase X's start is delayed to the end of the seventeenth week, there is an increase in project duration of 7 weeks.

On the other hand, if phase Y is forced to follow phase X, the increase in project duration is 1 week, shown as D_{XY} in Figure 8.17. Phase X's EFT

Project Scheduling with Limited Resources

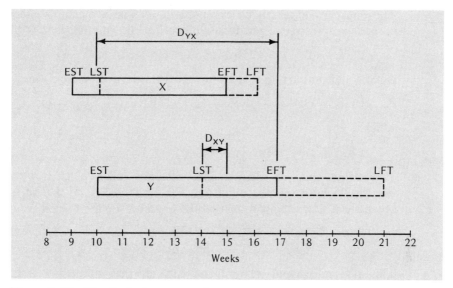

Figure 8:17 The limited resource rule

is at the end of the fifteenth week. Phase Y's LST is at the end of the fourteenth week. Thus delaying it to the end of week 15 increases the project duration by a week. Therefore, the optimal choice is to force phase Y to follow phase X, resulting in an increase in project duration of 1 week.

In general, the increase in project duration resulting from forcing phase j to following phase i will be given as follows.

$$D_{ij} = EFT_i - LST_j$$

If the difference between EFT_i and LST_j is negative, the increase in project duration, D_{ji}, is actually equal to zero.

Thus, when faced with the alternative of letting phase i follow phase j, or letting phase j follow phase i, one should choose the alternative that increases project duration the least. For example, assume the following is true.

$$EFT_i - LST_j \leq EFT_j - LST_i$$

Given this situation, then forcing phase j to follow phase i is the optimal decision. Let us return to the phases shown in Figure 8.17. It was decided that phase Y should follow phase X. This is borne out by the following calculation.

$$EFT_X - LST_Y \leq EFT_Y - LST_X$$
$$15 - 14 \leq 17 - 10$$
$$1 \leq 7$$

The optimal choice can be expressed in yet another manner. Assume it is optimal to force phase j to follow phase i. Thus, the following is true:

$$EFT_i - LST_j \leq EFT_j - LST_i$$

Adding and subtracting equal terms to each side, and manipulating the LSTs,

$$EFT_i - DUR_i + DUR_i + LST_i \leq EFT_j - DUR_j + DUR_j + LST_j$$

which is equivalent to

$$EST_i + LFT_i \leq EST_j + LFT_j$$

Thus, when choosing whether phase i should follow phase j, or vice versa, one should force the phase that has the largest EST + LFT to follow the other phase. For example, for the two phases in Figure 8.17, this sum is

Phase X: EST + LFT = 9 + 16 = 25
Phase Y: EST + LFT = 10 + 21 = 31

Thus phase Y should follow phase X. Note that although this calculation is correct, it does not clearly indicate the increase in project duration for the alternatives. It is often beneficial to know the increase in project duration precisely. Thus, comparing the EFTs and LSTs is often preferred.

If several phases conflict at the same time, the RSM procedure remains unchanged. However, it only focuses on two phases at a time. The RSM picks two phases that allow for the minimum increase (smallest positive or largest negative value of the difference between one phase's EFT and the other's LST) in project duration.

Let us now return to the project network described in Figure 8.16. A resource conflict was identified on week 1 of the project among phases A, B, and C. By inspection of Figure 8.16, or by CPM calculations, the EFTs and LSTs of the conflicting phases are determined to be

	EFT	LST
A	9	6
B	8	0
C	6	7

There are several logic changes possible. Phase A could follow phase B, phase B could follow phase A, phase C could follow phase A, and so on. However, the optimal choice (keep project duration to a minimum) can easily be found by comparing the phase EFTs and LSTs. Observe that the smallest number obtained by subtracting the LST of one phase from the EFT of another is $6 - 6$, or 0. Thus, the optimal choice is to let phase A follow phase C, with an increase in project duration D_{CA} equal to 0. This new project logic is shown in Figure 8.18.

Project Scheduling with Limited Resources

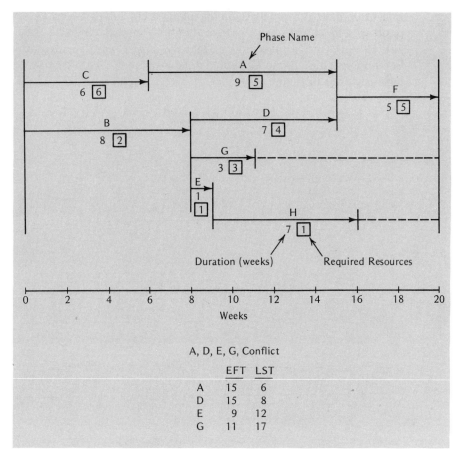

Figure 8:18 Rescheduling activity A

By progressing from left to right in the EST project plan in Figure 8.18, the next resource conflict is found to occur on week 9 of the project. Phases A, D, E, and G require a total of 13 resources versus an availability of 9. The EFTs and LSTs of these activities are as follows:

	EFT	LST
A	15	6
D	15	8
E	9	12
G	11	17

The minimum EFT of one phase minus the LST of another phase is found to be $EFT_E - LST_G$, which equals -8, implying an increase in project duration of zero. It is true that forcing phase E to follow phase G would also result in a zero increase in project duration, in that $EFT_G - LST_E$ equals -1. However, RSM rules indicate that the largest negative alternative should be chosen. Thus, phase G should follow phase E. This change of project logic is shown in Figure 8.19.

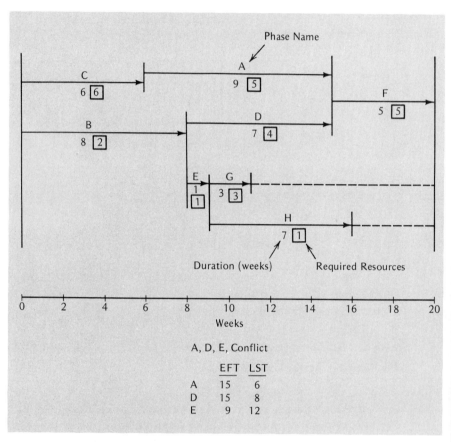

Figure 8:19 Relieving conflicts during week 9

However, the resource conflict at week 9 has not been totally resolved. As shown in Figure 8.19, phases A, D, and E still have a resource conflict. Their EFTs and LSTs are

	EFT	LST
A	15	6
D	15	8
E	9	12

Project Scheduling with Limited Resources

Of the possible combinations, $EFT_E - LST_D$ yields the minimum increase in project duration. The increase in project duration is given by 9 (the EFT of phase E) -8 (the LST of phase D), or an increase of 1 week. Thus, phase D should follow phase E, as shown in Figure 8.20.

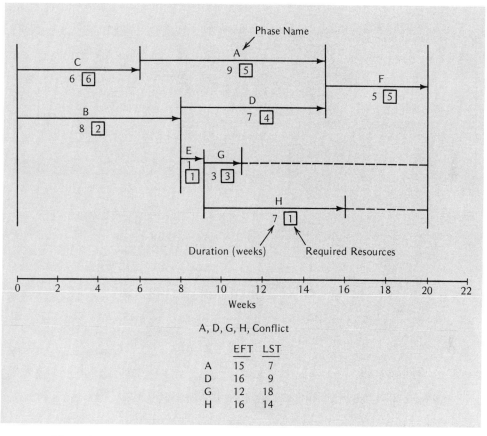

Figure 8:20 The third iteration of the RSM schedule

The resource conflict at week 9 is now resolved. However, on week 10, phases A, D, G, and H require a total of 13 resources versus an availability of 9. The EFTs and LSTs of these phases are

	EFT	LST
A	15	7
D	16	9
G	12	18
H	16	14

Forcing phase G to follow phase A does not result in increased project duration and satisfies the RSM requirement of choosing the largest difference between EFTs and LSTs. Thus, phase G should follow phase A, as shown in Figure 8.21.

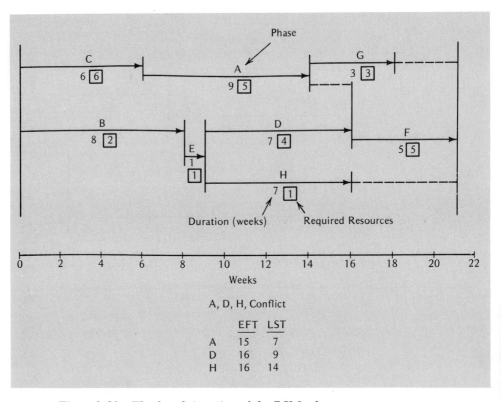

Figure 8:21 The fourth iteration of the RSM rule

Phases A, D, and H still have a resource conflict on week 10, since they require a total of 10 resources. Their EFTs and LSTs are

	EFT	LST
A	15	7
D	16	9
H	16	14

All alternatives yield an increase in project duration. The minimum increase in project duration occurs by forcing phase H to follow phase A. This is calculated as $EFT_A - LST_H$, or an increase of 1 week. This change of logic is shown in Figure 8.22.

Resource Leveling

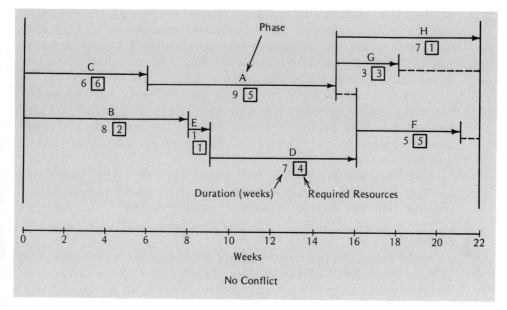

Figure 8:22 All the resource conflicts relieved using the RSM rule

Upon inspecting the project plan in Figure 8.22, no resource conflicts are found. Thus the plan shown in the figure has resolved the resource conflicts of the project's activities by increasing the project duration to 22 weeks. There are many plans that could be constructed to relieve the resource conflicts. However, many of these plans would produce a project duration in excess of 22 weeks. Thus, the RSM method is useful in determining an acceptable minimum duration project plan.

The RSM model takes advantage of possible activity delay times (floats) when determining the optimal logic between two phases by recognizing that a phase can start as late as its latest start time. The construction manager has the availability of the project phase floats to manipulate and control a project's time and cost.

RESOURCE LEVELING

The construction manager is not always constrained by limited resources when managing a project. For example, the construction manager can often consider various types of labor to be unlimited in supply, since available labor at the local union hall may far exceed any demand for labor for the project.

When unlimited project resources are available, it is possible to build the project in a duration determined by the technological constraints of the project phases. In fact, when the construction manager has access to unlimited project resources, he or she can manage the project using any of the many project schedules that can be constructed consistent with such a duration. This includes the earliest start time schedule and the latest start time schedule for the project. Many other schedules can be constructed by shifting phases within their allowable float times.

Let us return to the arrow notation plan in Figure 8.12. An earliest start time schedule for the project is shown in Figure 8.13, and a latest start time schedule in Figure 8.14. As shown in each of the figures, the amount of resource required for each schedule is not constant, but varies over the project's duration.

The variation in either the EST or LST schedule for the project may result in the schedule being less than optimal in regard to project time and cost. The fact that the amount of resource required varies with time can result in added project cost, lower productivity, or lack of project continuity.

The optimal project resource plan is a function of the overall resource allocation objective. Naturally, such an objective should be consistent with the construction manager's project objective. It is difficult to suggest an all-encompassing optimal resource allocation plan. The type of project and the construction manager's opportunities dictate the optimal plan.

Given a resource schedule such as one of those shown in Figure 8.23, and given the technological phase constraints, how does the construction manager develop an optimal project and resource schedule? By shifting phases within their allowable floats, the construction manager can develop such a schedule. In this section, the project duration will be assumed fixed as determined by the technological phase constraints. The fact that the construction manager has unlimited resources available allows the project to be built in such a duration. Admittedly, the construction manager can decide on a project duration in excess of this. However, such a schedule would provide more float for the resources. Thus, let us assume the duration of the project is actually equal to that determined by the technological phase logic. If an acceptable resource schedule can be developed for such a project duration, one can surely be developed for a schedule with a larger duration.

The ability to shift a phase within its allowable float time results in being able to vary the amount of resources from one time period to the next. Owing to the fixed nature of the resource requirement of each phase, and because of their logic and continuous nature, the construction manager will seldom be able to shift phases within the project plan so that resource allocation objectives are perfectly satisfied. For example, assume a construction manager has a resource allocation objective as shown by the solid lines in Figure 8.24. He or she may only be able to approximate such an objective by means of the resource allocation schedule shown by the dashed lines. However, such a resource schedule may be a vast improvement over the schedule dictated by only the technological constraints of the phases.

Resource Leveling

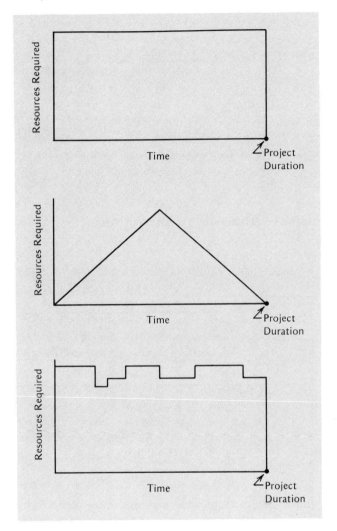

Figure 8:23 Different leveling objectives

Consider the project previously discussed and shown in Figure 8.12. Let us assume that a construction manager has an objective of minimizing the amount of resources required at any given time during the project. By referring to Figures 8.13 and 8.14, the maximums required by the EST and LST schedules are found to be 13 and 16, respectively. Perhaps the construction manager can reduce the maximum below the required 13. Let us attempt to accomplish this by shifting phases within their float times.

Since the project duration is determined by the critical path, the critical phases have no float time and therefore cannot be shifted. Thus, phases B,

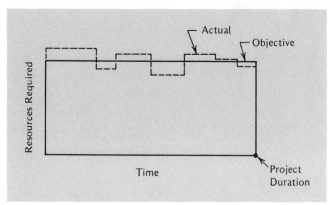

Figure 8:24 Actual versus planned objectives

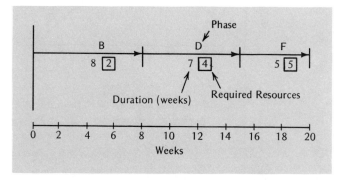

Figure 8:25 Scheduling the critical path

D, and F are performed as determined by their technological phase constraints. They are shown on a time scale in Figure 8.25.

Consider phases A and C. Both have float and rather large resource requirements of 5 and 6, respectively. It seems logical that to keep the amount of resource required at any time to a minimum, they should not be performed simultaneously. Thus, it becomes a matter of which phase should follow the other (delaying both phases accomplishes nothing). If phase C were to follow phase A, the project duration would be increased to 22 weeks. This is not allowable. However, forcing phase A to follow phase C does not affect the 20-week project duration. Thus, this logic change is made and is shown in Figure 8.26. Observe that, at this time, the project requires a maximum of 9 resources. If phases A and C occurred simultaneously, the project would have required a minimum of 13 resources.

Since phase G has a large float (a total float of 9 weeks) the construction

Resource Leveling

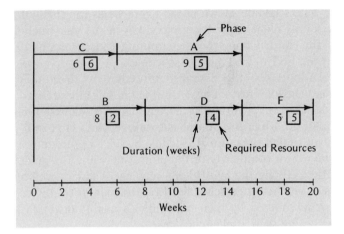

Figure 8:26 Scheduling activities A and C

manager has a large amount of freedom in scheduling it. Let us turn to the scheduling of phases E and H. Both phases have float and a resource requirement of 1. Regardless of when phase H is scheduled within its allowable time, it overlaps with phases A and D, forcing the maximum project required resource to 10. This is also true of phase E. Thus phases E and H are scheduled as shown in Figure 8.27. Finally, phase G is scheduled as in Figure 8.27 so that is does not conflict with the maximum required resource.

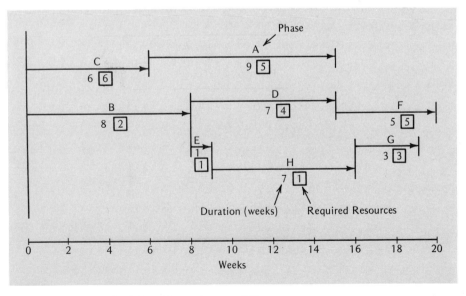

Figure 8:27 Schedule with maximum of 10 resources

The schedule shown in the figure yields the smallest maximum amount of resource required by the project. Given the construction manager's objective, the construction schedule requiring a maximum of 10 resources is better than the EST or LST project schedules.

Admittedly, the described procedure of determining the optimal resource schedule was not given so it could be computerized or even quantified into a set of rules. Although the optimal schedule was found for the problem in question, when the project network becomes large and complex it may be difficult to find the optimal schedule by such a procedure. However, regardless of the size or complexity of the project network, the guidelines discussed can greatly facilitate the construction manager's task of determining the optimal project and resource schedule. In particular, focusing on the phases in an order determined by their criticality (available float) facilitates the task of determining such a schedule.

EXERCISE 8.1

CPM CALCULATIONS

Given the information in the accompanying table regarding individual work packages to be awarded to contractors, draw the CPM diagram and determine the EST, EFT, LST, LFT, and TF for each of the work packages.

Work Package	Immediate Following Package	Duration (months)
A	B,C	3
B	I	2
C	G	6
D	E,F	2
E	H	7
F	I	5
G	I	9
H	I	10
I	—	1

Exercise

EXERCISE 8.2

CASH-FLOW ANALYSIS AND CPM

Assume that a construction manager awards each of the work packages listed in exercise 8.1 to contractors for the following amounts:

    A—$80,000   E —$ 70,000
    B —$60,000   F —$100,000
    C—$90,000   G—$ 45,000
    D—$50,000   H—$115,000
                       I —$ 40,000

Further assume that the construction manager's client wants to minimize the cumulative payments made to contractors through the tenth month of the project but also wants the project completed within 20 months. Assuming that the project owner is to pay contractors on a percentage progress basis and is to make payments during the month work is performed, calculate the cumulative cash payouts through the tenth month for both the earliest start time and latest start time schedules.

9

Phasing and Packaging Work

THE IMPORTANCE OF PHASING AND PACKAGING IN THE CM PROCESS

Two of the greatest potential benefits associated with the CM process are the phasing and packaging of work. These services are also two of the most frequently used. In fact, because of the common use of phasing and packaging in the CM process, some individuals think of phasing and packaging as the CM process. This is incorrect. Phasing and packaging can be part of the CM process; however, the CM process can also be performed without them.

Phasing is also referred to as *fast-tracking*. By either name, phasing is an overlapping of various parts of the design and construction phases of a project with the objective of condensing the overall project time schedule. Since time is money to a potential project owner, the phasing has the objective of decreasing the cost of a project. It may be possible to phase or overlap several parts or components of the project's design and construction. For example, it could be possible to let the construction contract for the substructure and the erection of the building frame before the design of the interior finishes is complete. Other examples of potential phasing or overlapping are discussed in this chapter.

Phasing is not without its potential disadvantages. Whenever some work is started before a related segment is designed, there is a distinct possibility of the need for change orders. Although these change orders can be minimized through an effective CM process, it follows that there is more potential for change orders associated with phasing relative to nonphasing.

Packaging refers to the awarding of separate contracts to several contractors to construct various elements or packages of a project. This contrasts to the general contracting process in which a single contractor, the general contractor, is engaged for the entire construction phase of a project. When the construction phase of a project is packaged, each contractor has a separate contract with either the project owner or the construction manager.

The objectives of packaging construction work include the reduction of project cost and the control or reduction of project duration. The cost reduction is attained by awarding packages compatible with obtaining competitive bids from contractors. The advocate of the CM process argues that it is possible to custom design the size and type of separate construction contracts to achieve a minimum construction cost. For example, if concrete contractors within a given geographic area are too small to bid on a certain project, the project owner, through the use of its CM firm, can choose to award three concrete contracts, each of which is of a size that enables the contractors to bid the work.

Packaging of construction work is often compatible with the phasing of work. Implicit with the phasing of work is awarding construction contracts as the design evolves. For example, the structural steel contract can be awarded as the electrical work is designed, the mechanical work awarded as the finish material work is designed, and so on. Consistent with phasing, an objective of packaging is to accommodate the continuity and possible reduction of the project schedule.

In practice, the number of construction packages awarded to contractors varies depending on several factors, including the complexity and size of the project. A range of six to as many as forty separate construction packages is typical for CM projects. Considerations regarding the number of construction work packages for a specific project, along with potential advantages and disadvantages of packaging are discussed in this chapter.

Project planning techniques like the critical path method discussed in chapter 8 can serve as the basis for performing both the phasing and packaging of work. The determination of what and when to phase and what to package should be based on the criticality, resource requirements, and timing of various work activities. The criticality of work activities, the determination of resource requirements as a function of time, and the occurrence of activities as a function of time are set out in a formalized project plan. The dependence of phasing and packaging on the project planning techniques discussed in chapter 8 are also discussed in this chapter.

PHASED CONSTRUCTION: THE BENEFITS AND DISADVANTAGES

Phased construction is also commonly known as fast-tracking or accelerated design and construction. Phased construction is simply the overlapping of the design and construction phases of a building process to shorten the total project time. In the traditional process, the construction project moves in a stage-by-stage manner throughout the entire duration (programming, design, design development, contract documents, bid and award of contracts, construction, and acceptance of completed project). This single-file nature takes considerable time since each phase waits for the preceding one's completion before it is initiated.

For almost every construction job the building site is determined early in the design process. Therefore, a contract for the rough site work can be let early in the building process. Similarly, any of the basic structural decisions can be made before all of the building details are established. This permits the early award of foundation and structural contracts. By using the CM process, construction can be initiated very early in the design process, instead of at the end of a lengthy design and contract preparation period. Weeks, months, and even years might be taken out of the traditional project schedule. Some proponents say that one of the fundamental facts of the phasing or fast-tracking process is that the greatest savings in time and cost in a construction program can be achieved during the design phase.

By using phased construction concepts, early phases of the project are put under contract while later stages are on the drawing board. Construction proceeds in the form of a series of small phased packages, each being contracted as its design is finalized.

Phasing is not a new concept, even though construction management ideas themselves are relatively new. It has been used on industrial construction for many years under the title of *phased construction*. It has been increasingly applied to other forms of building and is growing in popularity for commercial and industrial projects, especially when a design-build firm is engaged. This method lends itself well to construction phasing since the firm both designs and constructs the project. Construction can be initiated before design is finished under this method. Since the firm is directly involved in the design, early portions of the project can be let before the final design is completed. The CM process is, however, the process most associated with phased construction.

Once the construction manager has been chosen, he or she starts work with the job's architect. The work between these two professionals must be closely integrated such that smooth, logical designs can be achieved. The architect will outline the building concept and will proceed with the design activities in the order of the building's construction.

It is during these early designs that the CM firm has to have an eye towards phased construction. The project is planned so that major segments of the work can be designed and released for construction in a logical order throughout the duration of the job. The subsystems of the building must be selected in the beginning so that the succeeding design phase can proceed without delay. In other words, while one phase of the job is being constructed, the following phase is being designed, and so on.

One of the first steps to any project is preparing the contract bid documents. Before these contract documents can be drawn by the designer, the owner, through its construction manager, must be consulted to identify the schedule and order of the phases to be collected for separate contracts.

Once the bid documents have been prepared, the construction manager selects the prime contractor for the first phase by either open bidding or negotiation. At this time the owner is directly involved with the architect and construction manager and must now enter into a direct contract with the prime contractor. This is true even though authority and administration of the contracts has been delegated to the construction manager.

The prime contractors are selected at various times during the construction phase, with the number of prime contractors selected for any one job varying. Examples of the phases and the contractors used might be the site work followed by the building foundations, structural steel, electrical, mechanical, paving, and finally landscaping.

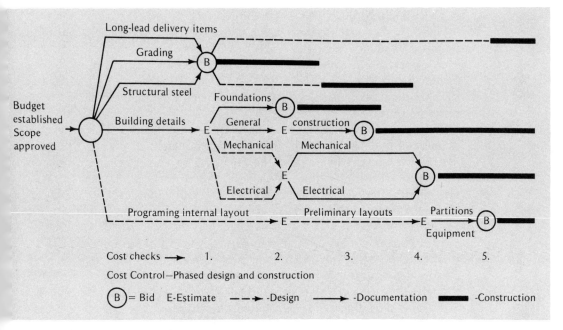

Figure 9:1 Benefits of phasing a project

The concept of phasing construction work and its relationship to reducing project duration is illustrated in Figure 9.1. The sequences of events for a phased construction project are contrasted to a nonphased project. The 24-week savings associated with the phasing of the work relative to the nonphasing should be viewed as only an example of the benefits of phasing. The magnitude of the benefits will vary with the project in question and the skill of the construction manager. Nonetheless, it follows that the overlapping of various project phases shown in Figure 9.1 will often lead to a condensing of the project schedule.

The types of projects that use phased construction the most are those that have relatively long construction durations. The longer the duration of a project the more the potential time savings.

Phasing or fast-track arrangements have been found to work well on commercial buildings and industrial buildings because of the predetermined requirements that these buildings have. Other projects might include hospital projects and governmental projects. Phasing can be taken a step further on a large, complex project by preparing flow sheets and process charts sequentially for different work areas, simulating several independent fast-track projects. If the work load is large enough, similar phases of work occurring in different areas could be let under separate contracts to different contractors.

It should be pointed out that the success of a phased construction project lies with the regularity with which the project is monitored and controlled by the construction manager. Control is discussed in the next chapter. Not every project needs to be fast-tracked. The reason for this is that all projects are not in the position to be helped by accelerated delivery. Factors such as size and duration can make phasing the project uneconomical for the benefits to be gained.

When speaking of the advantages that phased construction offers, the assumption is made that the building's design is integrated with the construction in the conceptual stage of the project. The major advantage of overlapping the activities is time. When the overall time requirements can be substantially compressed so that design and construction activities are occurring concurrently, the result is an accelerated delivery of the building to the owner.

At the same time, fast-tracking or phasing acts as a form of cost control. It is economically advantageous in an escalating market to start and complete construction quickly. When alternative design approaches are costed and evaluated early, a head start can be had on rising material and labor costs. The sooner a contractor can get started, the less affected he or she is by the rising job costs. Another advantage of phasing is that the analysis of needed materials can be completed during the design phase so that long-lead items can be ordered before a contractor bids the work. This saves time in that the contractor will only need to install the material when it arrives, the ordering having been done by the construction manager.

With a fluctuating market, phased construction provides the flexibility of packaging the prime contractors so as to take advantage of present conditions. If a certain phase of work involves material or labor that is currently undergoing a change in the market, adjustments can be made to let the condition stabilize.

Along with the advantages of phasing work, there are accompanying disadvantages. First, phased construction is extremely dangerous if rigorous schedules, accurate cost projects, and thorough analysis of each component of the project are not employed. The end result is serious confusion, delayed construction, and increased costs.

Another problem with phasing a construction job is that the owner does not know actual costs until the final prime contract is let, with the budget estimate only as good as the construction manager. The project owner is also encumbered with many contracts, and although he or she has delegated supervision of them to a construction manager, the ultimate decision remains the owner's. And, as mentioned before, if the architect's design is not well integrated, phased construction can result in costly early decisions, the consequences of which are not known until it is too late.

PACKAGING WORK: THE BENEFITS AND THE DISADVANTAGES

In the non-CM process, the project owner typically awards a single construction contract. The recipient of this contract is referred to as the general contractor. The general contractor in turn determines what work it chooses to subcontract. The general contractor also has the responsibility for choosing and contracting with each subcontractor. Usually the general contractor's choice of subcontractors is made with the objective of maximizing its own profit on the project.

It can be argued that the subcontractors are engaged to maximize the general contractor's profits instead of satisfying the project owner's time, cost, and quality objectives. For example, a general contractor might place all of its subcontractor selection criteria on low bids, independent of the subcontractor's willingness to adhere to schedules or work in harmony with other subcontractors. This criterion can work to the disadvantage of the project owner should the subcontractor disrupt the project schedule or subcontractor harmony, or perform poor quality work.

Should a subcontractor not perform in regard to schedule or quality, the project owner in the general contracting process is in an awkward position because of its lack of direct (privity) contract with the subcontractors. Although the project owner has control of the subcontractors through its general contractor, the fact that the project owner did not directly engage the subcontractors and that it does not directly make payment to the subcontractors can complicate or constrain a good working relationship.

Also implicit with the general contracting, non-CM process, is the awarding of a large dollar contract to the general contractor. If the construction work for a project is valued at $4 million, that contract amount is awarded to a general contractor. Often the general contractor will have to provide a performance bond for the entire project. The bonding capacity of a firm is almost totally dependent on the firm's financial strength with particular emphasis on the firm's equity. The result is that a small construction firm may not be able to bid work as a general contractor for relatively large cost projects.

In the CM process, the construction manager often replaces the general contractor and performs some of the duties of the general contractor. For example, the construction manager typically serves its client by determining what segments or packages of the planned construction work are to be awarded separately. However, it usually does not engage the contractors directly, as the general contractor does. Instead, the construction manager serves as an adviser to the project owner in recommending which contractors to consider and even which contractors should be awarded contracts. In the most common CM process, the contract agreements are made between the project owner and the individual contractors.

The construction manager concentrates totally on serving its client's project objectives in determining the dollar amount, type, and scope of contracts to award, along with its recommendations as what contractors to engage. The contractors are ideally selected based on their ability to deliver time, cost, and quality construction work to the project owner.

The packaging of several individual construction contracts provides several benefits relative to engaging a general contractor who in turn engages subcontractors. Included in these are the following:

1. Gives project owner more control over the individual contractors
2. Gives the project owner and its construction manager more flexibility and control in the selection of the individual contractors
3. Allows for a reduction in the amount of overhead related to one contractor marking up its subcontractor's work
4. Provides for a means of awarding contractor work packages that are of a scope and size compatible with procuring competitive contractor bids
5. Provides for a means of structuring work packages that accommodate the satisfying of various governmental or external agency requirements, including minority set-aside programs
6. Enables a control system that pinpoints individual contractor performance or nonperformance with the potential for improved control of the construction phase
7. Accommodates the phasing of construction work in that individual construction packages can be awarded as a design evolves

As noted earlier, the mere fact that the project owner has a direct contract relationship with a contractor infers an improved capability to control or monitor a subcontractor. Naturally the terms of the contract agreement between the project owner and the individual project contractors dictate the working relationship and control potential. However, at a minimum this control should entail the owner's insistence on the contractor's adherence to a project time schedule.

Although the project owner's direct or privity of contract with individual contractors should improve the owner's time, cost, and quality control potential, the individual contractors can also benefit from this direct contractual relationship. For one, any grievance a contractor has can be handled directly with the owner or its construction manager. This contrasts with the subcontractor's having to deal with a middleman, the general contractor, in attempting to resolve a subcontractor/project owner dispute. When separate individual construction contracts are awarded to individual contractors it is also possible for the individual contractors to get paid more quickly and to receive their retainers sooner. Because there is no middleman, and because of the delineation of individual work packages, contractor-held retainers can be released as an individual contractor completes its work packages. This contrasts to the situation in which an early-phase subcontractor does not receive its retainer until the project is completed and the retainer is released to the general contractor.

The fact that individual contractors are engaged directly by the project owner (through the assistance and advice of the construction manager) enables the project owner to be selective in its choice of contractors. In nonpublic construction projects, the project owner can limit its invitation to bid to contractors it judges appropriate. In addition, the project owner can choose to award a contract to a contractor other than the low bidder if the owner judges the low bidder to be less conscious of project schedules and quality construction work than another bidder. In effect, the project owner, through its construction manager, can pick and choose the contractors to be awarded work. This contrasts to the nonpackaging process in which the general contractor chooses many of the contractors to be awarded construction work.

The project owner's ability to pick and choose its contractors diminishes when the project is public construction. Various governmental regulations can result in a need for the project owner to conduct an "open" advertisement for contractors and in turn to award contracts to the low bidders. However, even on government projects, the project owner can control the bidder's invitation list and in certain instances be able to award work to a non-low bidder.

One of the benefits of packaging work that is not frequently cited is the potential reduction of contractor markup associated with one contractor working for another contractor. Whenever construction work is packaged, there is a flattening out of the organizational structure relative to the nonpackaging process. The more subcontractor relationships that exist, the more doubling up of contractor overhead and profit an owner must incur.

For example, for a $1 million project, let us assume that individual contracts are packaged and the individual contractors mark up their work 20 percent for overhead and profit. The result is that the project owner incurs $200,000 of overhead and profit expenditures.

Instead, let us assume that a general contractor is engaged and subcontracts 50 percent of the work. Assume the subcontractors mark up this work 20 percent, or $100,000. Further assume that the general contractor marks up its own work 20 percent and the work of its subcontractors 10 percent. The result is that the general contractor has a markup of $110,000 ($100,000 from its own work and 10 percent of $100,000 or $10,000 on the subcontractor work). The result is $10,000 of added contractor markup for overhead and profit relative to the packaging process. This added cost is passed on to the project owner.

Perhaps the most important benefit associated with packaging construction contracts is the potential for the project owner to impact the dollar amount of the construction contracts awarded favorably. The dollar size and scope of construction work packages dictate which contractors will bid and to a degree how much they will bid. For example, if work package contracts are relatively large in dollar amount, contractors possessing relatively few resources and only marginal financial equity will not be able to bid the work. Similarly, if the bid packages are small in dollar amount, large contracting firms or "out-of-town" contracting firms may not be interested in bidding.

It is also true that within a given geographic area, certain types or sizes of contractors may be in need of work or may already be busy. Depending on the contractors' positions at a specific point in time they may be willing or unwilling to bid at a competitive or fair price for a project. Knowing the particular contractor's status can enable the construction manager to shape the contracts to accommodate the receipt of competitive and fair contractor bids.

The competitiveness of contractor bids is not only dependent on the dollar size of the proposed packages. The type or scope of work included in specific packages also dictates the competitiveness of the contractor bids. For example, a specific contractor might be interested in bidding a package that includes both the mechanical and electrical work. The same contractor might not be willing to bid either the mechanical or electrical work if awarded separately.

The ability of the construction manager to set out competitive work packages is dependent on the construction manager's estimating skills. The estimating skills and techniques discussed in chapter 6 provide the base for determination of competitive bid packages. For example, the ability to estimate the cost of three separate concrete work packages for a project relative to estimating the cost of one package that includes the work for the three packages is fundamental to the determination of work packages to be used.

In addition to the use of packaging to control the cost of construction by shaping the contract work packages to the contractor marketplace, the packaging of work can be used to satisfy other contractor selection objectives. For example, if the project uses government funds, a certain percentage of the construction work may need to be awarded to minority contractors. The size and scope of the packaged contracts can assist in satisfying this requirement.

Assume all minority firms within a given geographic area are small and can only handle small contracts. Let us further assume that the project owner determines that several minority painting contractors work in the area. The letting of several small painting contracts can enhance the interest of the minority firms and result in the project owner being able to satisfy a government-imposed or owner-imposed minority set-aside program. Similarly, an owner's interest in limiting the awarding of construction contracts to in-town contractors can be accommodated by defining the construction work packages to be compatible with the characteristics of the in-town firms.

The packaging of work can also enhance the project owner and construction manager's control capability over the individual contractors. A concept in accounting, referred to as *responsibility accounting*, indicates that the ability to control a cost or cost object is dependent on the ability to identify expected performance for the cost or cost object and the ability to monitor the individual cost or cost object. The concept infers that if a cost object cannot be isolated, budgeted, and monitored, it cannot be effectively controlled.

The setting out of clear individual construction work packages to contractors should result in a better measurement system for control of the individual contractors relative to the general contractor process. It can be argued that when several subcontractors are engaged indirectly by a project owner, the inefficiencies (or efficiencies) of one of the subcontractors are clouded by the existence of other subcontractors. In effect, the project owner can lose control of the entire construction team because of its inability to pinpoint good or poor performance of individual contractors. The packaging of work, along with well-defined, concise, and thorough individual package contracts should alleviate this difficulty.

As noted earlier in the chapter, the very process of phasing or fast-tracking may be impossible without the use of separately awarded construction packages. Phasing implies the awarding of construction packages as the design and construction proceed. Phasing is impossible or ineffective without packaging individual contracts. To this end packaging can be viewed as a means of phasing and/or condensing the construction schedule.

Although the primary objective of packaging is centered on minimizing the project cost by means of securing competitive contractor bids, packaging has to be performed with an eye to enhancing of the continuity of the project schedule. For example, although two separate structural steel contracts for a multibuilding complex might yield lower bids over a single structural steel contract, the engagement of two contractors can result in added construction time. Both time and cost considerations should be part of the determination of packages to be awarded.

The number of distinct construction packages to be established depends on several factors, including the complexity of the construction project, the dollar size of the project, amount of phasing to be performed, and the contractor marketplace at the time the project is let. The number and type of packages awarded often are established consistent with the types of construction work performed by individual contractors. For a building construction project the packaging is done consistent with the lists of trades shown in Figure 9.2. The list of trades serves as a natural breakout of work. This does not imply that the construction manager will always award a contract for each of the trades shown in Figure 9.2. For example, knowing the masonry contractor marketplace in a given area, the construction manager can choose to award a single contract to a masonry contractor for both exterior and interior masonry instead of two separate masonry contracts.

General conditions	Curtain wall
Sitework	Lath and plaster
Utilities	Drywall
Roads and walks	Tile work
Landscaping	Terrazzo
Excavation	Accoustical ceiling
Foundation	Resilient flooring
Caissons and pilings	Carpet
Formed concrete	Painting
Exterior masonry	Toilet partitions
Interior masonry	Venetian blinds
Structural steel	Special equipment
Miscellaneous metals	Elevators
Ornamental metal	Plumbing
Carpentry	Sprinklers
Waterproofing and dampproofing	HVAC
	Electrical
Roofing and flashing	Miscellaneous trades
Metal doors and frames	Parking
Metal windows	
Wood doors, windows	
Hardware	
Glass and glazing	

Figure 9:2 Packaging work by trades

Although the construction manager can award close to forty contracts similar to those listed in Figure 9.2, it may also choose to award as few as six, as shown in Figure 9.3. These packages are compatible with a systems breakout for a project. In effect, each package is a building system. Individual contracting firms are often structured to perform construction work for one of the systems. Each of the packaged systems can be phased to yield a shortened project construction schedule relative to a linear, nonpackaging approach.

> Foundation System Contract
> Building Frame System Contract
> Finishes System Contract
> Electrical System Contract
> Plumbing System Contract
> Mechanical System Contract

Figure 9:3 Packaging work by systems

Packaging of construction work is not without its potential disadvantages. These disadvantages must be weighed when evaluating the use of packaging and the number of packages to be awarded. Included in the disadvantages of packaging construction work are the following:

1. The potential of omissions in covering all work to be performed because of the separation of the total construction phase work into pieces
2. The potential for overlaps in contractor responsibilities because of several construction packages
3. Loss of contractor coordination and control because of the lack of a single contractor
4. Less bidding competition among specialty contractors because of the need for the individual contractors to bond their own portion of the construction work
5. Potential increased contractor costs associated with managing individual contracts

Although each of these disadvantages should be recognized, it is also true that an effective construction manager can reduce or eliminate these negative considerations.

The first potential disadvantage is the most serious in regard to the objectives of delivering a project within its time and cost objectives. The more the construction phase of a project is broken into individual packages, the more likelihood there is that needed construction work or responsibilities will be omitted in defining the contract packages. For example, various segments of project cleanup work, normally performed by the general contractor, may be unintentionally omitted in the contract packages.

The unintentional omission of construction work or responsibilities in the defined construction packages has at least two negative effects on the time and cost of a project. For one, the omission of work will result in the need for the project owner to initiate costy change orders. More often than not, the cost of change-order work is greater than if the work had been included in the original contract documents. In addition, a change order usually results in added project time.

Poorly defined construction packages that omit necessary work also result in some contractors not bidding. Contractors can usually detect when contract packages are going to result in problems. Fearing that they will have to perform the omitted work at no added compensation, a contractor may decide not to bid the work. Although it is true that a change order can be initiated, a contractor can never be sure of this. In many instances a contractor simply has to "eat" the cost of doing the omitted work. A lack of contractor bids is often an indication of poorly defined construction packages. Given this fact the project owner is penalized by packaging due to its inability to obtain competitive bids. In the general contracting, nonpackaging approach, the general contractor can take the responsibility for covering omitted work.

Just as defined construction work packages can omit needed work or areas of responsibility, the packages can include redundancies or overlaps of work or responsibilities. For example, various segments of the mechanical contract can also be included in the electrical contract. The result of contract redundancies or overlaps will lead to disputes or confusion that can result in added project time or cost, construction claims against the project owner, and lawsuits. Any of these conditions works to the disadvantage of the project owner as well as the contractor team.

The question of the improved or decreased contractor communications or control associated with packaging relative to a general contracting nonpackaging process is debatable. As noted in the previous section of this chapter, the privity of contract between the project owner and individual contractors gives the project owner more potential communication and control over the contractors relative to the communication and control the project owner has in dealing with subcontractors engaged by a general contractor.

Splitting the construction phase into several contracts can also result in less communication and control because of the lack of a single contractor coordinating the entire construction team. Without this coordination, each of the contractors performing packaged work lacks direction and in effect "the right hand does not know what the left hand is doing." Obviously this lack of communication can be alleviated by the construction manager's coordination effort. The construction manager provides the same coordination (actually the construction manager should attempt for improved coordination) that the general contractor provides in the nonpackaging approach. Without this coordination, it follows that the packaging of construction work can lead to less construction phase communication and control.

The project owner often requires that performance bonds be secured for the construction phase of a project to ensure that the project will be built according to specifications and to protect against contractor financial failure. When construction work is packaged, any bonding requirements are typically delegated to the individual contractors.

In a general contracting, nonpackaging approach, the general contractor can provide the bond for its subcontractors. In effect each of the specialty contractors is permitted to perform work without the need to secure a performance bond. The requirement that individual contractors provide a bond for their work can lessen competition because some contractors will not be able to obtain performance bonds. This consideration, although often cited as a disadvantage of packaging and the CM process, is overemphasized. The actual number of contractors eliminated from bidding a project because of inability to secure a bond is only marginal. It also follows that if a contractor lacks sufficient financial strength to secure a bond, then the project owner may be avoiding subsequent problems by deletion of the contractor from its bidding consideration.

Finally, a disadvantage of the packaging process is that splitting the construction work into several packages results in added contractor learning and managing costs. It is true that splitting work into several contracts does place more responsibility on the individual contractors to manage their own segment of work. It follows that the need for each contractor to manage itself results in a total aggregate learning or managing cost in excess of the implied project learning or managing cost of the general contractor. However, this is difficult to prove or quantify. It also is true that the very argument that the learning or managing time is increased is offset or defeated if the construction manager does an effective job of providing leadership, coordination, and communication for the individual contractors engaged.

THE ROLE OF THE CONSTRUCTION MANAGER IN PHASING AND PACKAGING

Without the construction management process and the project owner's engagement of a construction manager, the phasing or packaging of work must be performed by the project owner or its architect/engineer designer. Many potential project owners lack the expertise needed to perform phasing or packaging effectively. Many project owners are one-time builders and as such do not have the everyday knowledge of the design or construction phase that accommodate phasing or packaging.

There are at least two difficulties associated with centering the responsibility for phasing or packaging with the architect/engineer design team. For one, experience has shown that the design firm's expertise centers on design and not on construction, the estimating of the cost of construction, or the various responsibilities or work processes of the construction contractors. These skills are needed to phase or package work effectively.

Even if a project's design team has the necessary skills, there is a question of soundness or even ethics of the design team making decisions that affect the choice of specific contractors. To a degree, the size and scope of the construction packages dictate which contractors will be engaged. Some architectural/engineering design firms do not feel that they should be involved in such roles. They see their role as the project designer. It has been long-standing tradition in the design profession that the design firm should not select materials, methods, or contractors. Not all designers would agree with this view. However, it is true that the design firm's involvement in the phasing and packaging of work can place it in a position to receive criticism from contractors and suppliers.

The construction manager is in the most independent role in regard to phasing or packaging work. Its independence puts it in an objective position that serves only its client. Perhaps more important than its independent role is that the skills of a construction manager are most compatible with the performance of these functions. As noted in chapter 4, the effective construction manager has both design and construction skills as well as management skills. To overlap the design and construction phases of a project, knowledge of both design and construction is basic.

In addition to knowing the design and construction process, the construction manager has cost estimating skills and knowledge that are important ingredients of phasing and packaging work. Knowing the cost impact of construction work assists in defining what work is to be phased and the size and scope of construction packages.

The construction manager's success in phasing and packaging is also dependent on its knowledge of the labor, material, and contractor market in the project's area. For example, the need to have the owner purchase construction material during the design phase is dependent on the availability of the material. This availability varies from one area to the next. It follows that a construction manager performing work in its own geographic location has an advantage over a nonlocal construction manager in regard to the performance of the phasing and packaging functions. However the nonlocal firm can overcome this disadvantage by learning the prevailing conditions in the area, and from past experiences and skills in performing the functions of phasing and packaging.

Although both phasing and packaging are possible apart from the construction management process and vice versa, they are usually accommodated by entry of the construction manager and construction management process. It is also true that the construction manager without the skills to phase and package work will often prove to be an ineffective construction manager.

THE ROLE OF PLANNING TECHNIQUES IN PHASING AND PACKAGING

The critical path method (CPM) of planning and scheduling presented in chapter 8 can serve as an effective basis for the phasing and packaging of work by the construction manager. The identification of what work to phase and what work to package and the impact of either phasing or packaging on the project's duration and cost can be determined from a formalized project plan such as the critical path method.

Consider the oversimplified CPM diagram for a project illustrated in Figure 9.4. Calculations performed indicate that activities B, F, G, H, and J are the critical activities that dictate the 39-week project duration. The activity letters are abbreviations for real activities, such as order structural steel, erect substructure, frame building, and so on.

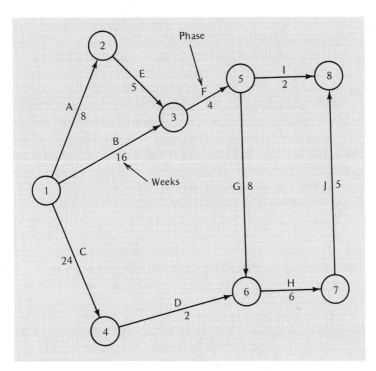

Figure 9:4 Using CPM for phasing and packaging

Let us assume that the 39-week schedule is not acceptable and that the construction manager must condense the schedule to 30 weeks. One means of shortening the project duration is through phasing. Packaging may be another means of reducing project duration.

If phasing is used to shorten the project, the question becomes what activities should be phased? Inspection of the CPM diagram in Figure 9.4 indicates that neither activity A nor E is on the project's critical path. Although it may be possible to start a part of activity E while activity A is finishing, the fact that these activities are not critical leads one to conclude that phasing these activities would not shorten the project's duration significantly. Instead, the construction manager could add cost to the project through ineffective phasing.

Although activities A and E are not likely candidates for phasing, the critical activities B, F, G, H, and J are activities that would yield time benefits if phased. For example, any overlapping of activities B and F would shorten the project duration. The amount of impact on the overall project duration that this phasing would have would depend on the degree of overlapping. The fact that F follows B constrains the amount of phasing possible. However, this does not mean that some overlapping is not possible. For example, activity B could refer to the forming and setting of scaffolding for concrete pours and F to the pouring of the concrete. Although the concrete cannot be poured until the forms are set and scaffolding is erected, some of the concrete can be poured as parts of the forming and scaffolding are erected.

Just as a prepared CPM plan can point to activities or work that should be phased, the plan can also draw attention to work that should be packaged and the impact of the packaging on the project duration. As is true of phasing, the activities that yield the greatest benefits from packaging are critical activities.

Let us assume again that the construction manager is attempting to shorten the project duration of 39 weeks for the project illustrated in Figure 9.4. Activity G is on the critical path and is scheduled to take 8 weeks to complete. Let us assume that the 8-week duration is based on the fact that a single contract is to be awarded and the equipment availability of any single contractor prohibits a duration of less than 8 weeks. In other words, the duration is dictated by the ability of a single contractor's financial or resource availability to perform the work.

Given the need to shorten the project duration, the construction manager can give thought to the awarding of multicontracts for activity G in Figure 9.4. For example, let us assume that the work is packaged for contract awards to three separate contractors. Although the financial or resource constraints of any single contractor prohibit a duration of less than 8 weeks, the combined financial and resources of the three contractors awarded parts of the work represented by activity G can provide a reduced duration. Perhaps the three contractors can perform work simultaneously with the result being that the duration is reduced from the 8 weeks. It is unlikely that the 8-week duration can be decreased by a factor of 3 merely because three contractors are engaged. A duration of 4 to 6 weeks is more likely. Nonetheless, activity

The Role of Planning Techniques in Phasing and Packaging

G as well as the overall project duration can be shortened through this packaging of work.

It is the CPM plan illustrated in Figure 9.4 that points the construction manager to the potential benefit of splitting or packaging activity G by indicating its critical nature. Similarly, the construction manager using the CPM plan illustrated in Figure 9.4 can easily determine that it may not be advantageous to attempt to package work represented by activities A or E. These activities are not critical. They have possible float time and have little impact on the overall project duration. Naturally should these activities become delayed in the design or construction phase of the project, they may become critical and at that time be candidates for phasing or packaging. However, as scheduled in Figure 9.4, the project's duration is sensitive to activities B, F, G, H, and J. It is these activities that should receive the attention of the construction manager. Naturally the cost impact as well as the duration impact from the phasing or packaging should be analyzed.

The application of a formalized project plan such as CPM to phasing or packaging is not limited to setting out the critical versus noncritical activities or work processes. The knowledge of the float or possible delay time can also assist in performing the phasing or packaging functions. For example, let us assume that a construction manager is considering packaging the work represented by activity C in Figure 9.4. The construction manager is considering awarding two contracts for this work with the objective of decreasing the cost. Based on the construction manager's estimates, the dollar amount of two contracts awarded for the work represented by activity C will sum to an amount less than if the work were awarded to a single contractor.

Although packaging two contracts for activity C results in a lessened construction bid cost, it does not always follow that the awarding of packages will shorten the activity's duration. In fact, it can happen that the overlap of functions and the need for two contractors to move in causes the duration of the work to increase. For example, let us assume that the construction manager estimates that the duration of activity C will be extended by 2 weeks, from 24 weeks to 26 weeks. The construction manager is now faced with a dilemma. The cost savings from packaging two contracts could be offset by an extended activity and perhaps project duration.

Having the CPM plan illustrated in Figure 9.4, the construction manager is in a position to evaluate the cost savings relative to the possible added duration. Inspection of Figure 9.4 and the performing of CPM calculations would yield the fact that activities C and D actually have 2 weeks of float time. This float time will allow for a 2-week extension of the work represented by activity C without any effect on the overall project duration. It follows that the construction manager can package the work for activity C, obtain a lower project cost, and not have any increase in project duration or cost. It is the knowledge of the existence of the floats generated in the CPM plan that enables this decision.

Other applications of a CPM plan and the accompanying activity floats relevant to phasing or packaging are dependent on the project characteristics and the creativity of the construction manager. However, without such a plan it is questionable if the construcion manager can effectively phase or package work to control project time or cost.

EXERCISE 9.1

POTENTIAL DIFFICULTIES OF PHASING WORK

Phasing construction work has the objective of shortening project duration and decreasing project cost. However, the phasing of work on some projects has actually led to increased project duration or cost. List some of the difficulties or potential problem areas that might lead to either an extended project duration or cost when a project is phased.

EXERCISE 9.2

PACKAGING CONSTRUCTION WORK

For the CPM diagram in Figure 9.4, assume that each of the work packages can be broken into several packages that will affect the cost and duration (in weeks) of each package. Assume the construction manager has analyzed the packages and has determined the cost and duration possibilities in the accompanying table. Given this information, determine 1) the minimum cost schedule, 2) the minimum duration schedule, and 3) the schedule that appears best in regard to both minimum cost and duration. For each schedule also set out how many packages are to be awarded for each work phase.

Exercise

Work Phase	1-Package Contract Sum Cost/Duration	2-Package Contracts Sum Cost/Duration	3-Package Contracts Sum Cost/Duration
A	$50,000/8	$60,000/6	$70,000/5
B	20,000/16	18,000/14	18,000/11
C	60,000/24	70,000/23	70,000/21
D	40,000/2	—	—
E	10,000/5	10,000/8	10,000/8
F	50,000/4	45,000/3	42,000/3
G	30,000/8	24,000/10	28,000/12
H	10,000/6	12,000/4	14,000/4
I	10,000/2	12,000/2	—
J	40,000/5	45,000/6	45,000/6

10

Controlling Project Time and Cost

THE IMPORTANCE OF CONTROL

The management functions necessary to deliver a construction project within defined time, cost, and quality objectives can all be viewed as part of two functions: planning and controlling. Planning involves the cost estimate prepared for a project and the determination of required resources, including the preparation of a time budget. Planning by itself does not ensure that a project will be delivered within the time, cost, and quality constraints. Planning enables these objectives to be accomplished. It is the control function that can bring planned objectives into reality.

Control can be defined as the timely comparison of actual performance versus planned performance with the objective of detecting inefficiencies and correcting them. Without the elements of this definition, control of project time, cost, and quality will fail. For example, if the function is not performed in time, the detection of an inefficiency can occur when it is too late to correct it. To control either the duration or cost of a construction project, control procedures need to be performed at least monthly and more likely weekly. For very short duration projects, control procedures may have to be performed daily.

The control function can be viewed as part of accounting. Accounting includes the use of forms, data collection, data processing, and various reports to measure performance. Given this fact, it becomes necessary to measure actual performance through the collection and processing of information. The forms and procedures used to accomplish this are referred to as the *cost control system.*

It should be noted that part of the definition of control is the comparison of actual versus plan. This is another argument that can be made for performing the formalized estimating and planning procedures described in earlier chapters. One might justifiably argue that without a plan one cannot control.

The comparison of actual versus planned performance does not by itself infer the control of time, cost, or quality. The comparison merely flags possible problem areas or inefficiencies. In this regard, the comparison serves as a detector of potential problems or inefficiencies. For example, the comparison may flag the fact that a mechanical contractor is falling behind schedule and could subsequently delay other contractors and the overall project schedule. Similarly, a calculation of a contractor's actual versus planned progress can indicate that the dollar amount of a contractor's payment request is in excess of the amount due the contractor from the project owner.

Control does not stop with the detection of a problem or inefficiency. The last and most important step is correction. Cost accounting data can flag a potential problem. However, the correction of it could be dependent on a more detailed analysis of the cost or process to be controlled. For example, a report might indicate that a contractor is behind schedule relative to planned progress. Further analysis performed at the job site may show that the contractor's work procedures are inefficient, including a poor balance of manpower and equipment or the use of non-productive technology. Analysis can also indicate that the contractor would be more effective if it pumped concrete rather than use a crane. Without this last step of analysis and correction, all of the previous control elements serve no purpose.

The construction manager focuses primarily on project time and cost in performing the control function. Although construction quality is relevant, the fact that the contract documents set out required quality reduces the likelihood of suboptimal quality.

In performing its control function, the construction manager is constrained by lack of direct contact with the contractors engaged. However, this does not lessen the need for CM control. This chapter discusses the control elements of a construction manager's project control system, including setting out some of the difficulties associated with the CM firm in regard to its noncontractual relationship with the contractors.

THE CM FIRM'S ROLE IN PROJECT CONTROL

One of the potential disadvantages of the CM process identified in chapter 2 was the fact that the construction manager may be in a position of responsibility without authority in regard to a project's construction contractors. If this is true, it proves difficult for the entire CM concept. High on the list of the benefits supposedly associated with the CM process is the control of project time and cost.

Assume a project contractor in a CM process is performing nonproductively. For example, assume the CM firm believes the contractor should employ eight carpenters rather than five to optimize productivity. Can the CM firm impose this change on the contractor? The answer to this is likely no. However, even if the CM firm had direct contract with the contractor as would the general contractor with its subcontractor, the answer would still be no. In other words, the problem of control in CM is little different from the problems associated with a general contractor controlling subcontractors.

One might ask why a construction manager should be concerned with the occurrence of a time or cost inefficiency of a contractor on its project. If a contractor has agreed to a contract amount and a completion date for its work, why then should the project owner or construction manager be concerned with the time or cost performance of the contractor. The reasons for the concern are twofold. One, a contractor's nonperformance will result in negative effects on other project contractors' time schedules, with the result that the owner will not receive the project according to contract time or cost conditions. Although the project owner can claim legal damages against the contractor due to the inefficiency, such a claim usually adds to the project time or cost. The best lawsuit is one that does not occur.

A second consideration relevant to why the construction manager should be concerned with a contractor's performance is that if the contractor is productive and does in fact make a fair profit on its work performed, it probably follows that the project owner also obtains its project within its time, cost, and quality objectives. It does not follow that in order for a project owner or construction manager to obtain its objectives, the contractors have to be squeezed to the point of not making a fair profit. The end result is that the construction manager should be concerned about contractor performance because the only means of delivering the construction phase to its client is through this performance.

Given the lack of direct contract and the fact that individual contractors are not the employees of the CM firm, how does the construction manager gain control over contractor performance? There are at least four actions the construction manager can take to ensure this control.

1. Select productive and "team-playing" contractors
2. Control the contractors through its authority as an agent of the owner
3. Manage or control the contractors through its ability to control owner payments from the project owner to the contractors
4. Control or manage the contractor through the authority it gains with the contractors by providing the contractors with management services

The best way to gain control over contractors is to engage contractors that do not need to be controlled. The construction manager's involvement in the selection of contractors can serve to ensure that productive and cooperative contractors are used. This is especially true when the project is private construction and the project owner can engage a firm other than the low bidder. The CM process works best when the contractor and CM firm cooperate, communicate, and respect each others skills and abilities. The contractor needs to respect the CM firm as a manager and the CM firm must respect the building skills of the contractor. If this is the case, the CM firm will be in a position to control the project time and cost effectively through its coordination and communication with the contractors.

The CM firm has the same authority over the contractors as the project owner, through its agency contract with the owner. This authority is only as effective as the willingness of the contractor team to be controlled. In other words, even with this authority, an unwilling or uncooperative contractor can reject control that is mandated. The same is true of the control related to the CM firm's holding back payment from a contractor. Control that is forced upon a contractor or control that is attained through pressure will prove ineffective by enhancing adverse relations.

The most effective control the CM firm can attain over the project's contractors is through gaining respect as an effective manager serving both the contractor team and client by being fair to each, and through leadership and management practices that assist both parties. The effective CM firm provides "free" management services to the contractors during the construction phase of the project. These services include scheduling, advice regarding construction methods, and serving as an overall troubleshooter. If the CM firm can assist the contractors in being productive, it follows that the contractors will be profitable and cooperate with the CM firm. The contractors are put in the role of builders and the CM firm provides the management leadership. If the CM firm can assist the contractors in their production objectives, the contractors will be more receptive to requests and control practices of the CM firm.

This helping of the contractors through providing management services and advice can benefit the project owner. Contractors that are aware of these CM firm services and favorable working environment are likely to bid work for the owner at a competitive fair price. Secondly, it usually follows that a contented and profitable contractor usually results in a project being delivered on or near a project owner's time and cost budget. In effect, it is possible for the project owner, construction manager, design team, and contractor all to attain their goals. The CM firm striving for this condition is likely to attain the most effective control.

The CM firm might be viewed as the "captain of the team" in regard to the control function. It serves as a decision maker and a sounding board. It coordinates and leads job-site construction meetings, detects inefficiencies through monitoring the work progress, serves as an arbitrator in owner-contractor disputes, and serves as a funnel and administrator of payments made to the contractors from the project owner. The CM firm must be aggressive and equitable in performing its control function. Means of accomplishing control are discussed in following sections.

There is a question as to whether the construction manager is a supervisor or observer of the construction phase. Supervisor implies responsibility for the performance of subordinates. Because the CM usually does not have direct contact with the project's contractors it is usually reluctant to be held liable for the performance of the contractors. It follows that the CM firm will contract with the project owner for observation of the construction phase of the project. The firm practices the same procedures regardless of whether it contracts for supervision or observation. However, because of the different liabilities associated with supervise versus observe, the firm will contract for observation.

The CM firm usually contracts to control a project's time, cost, and quality in both the design and construction phases. In effect, the control practices it imposes in the construction phase need to be executed over the design team in the preconstruction phase. However, the CM firm's emphasis in the design phase is more centered on planning and advising, whereas control is more important in the construction phase. The importance of control in the construction phase is highlighted by the fact that there are more entities involved in the construction phase and that this phase of a project is subject to the most uncertainties.

COORDINATION AND COMMUNICATION MEETINGS

Much of a construction manager's success or nonsuccess relevant to delivering a project on time and on budget is dependent on its ability to coordinate the entities involved in the project's phases. Its ability to coordinate these entities is in turn dependent on the communication that occurs in the meetings that are necessary during every phase of a project. Included in these meetings are owner/construction manager meetings, design team/construction manager meetings, prebid meetings, and daily, weekly, or monthly job-site meetings with contractors.

The construction manager is the captain or leader of these meetings. The means the CM uses to conduct these meetings will often result in their success or failure. Not all meetings are necessary or successful in accomplishing their objectives. The result is that the construction manager, as part of the performance of its contracted control function, should be attentive to the skills and procedures conducive to the holding of successful meetings. Successful meetings do not just happen, they are planned.

A meeting is the fastest and surest way of passing information to a group of individuals. It can save the construction manager a lot of time that is wasted in sending and answering numerous memos and letters. A meeting can reduce tensions and resolve conflicts among individuals. Perhaps more important, the meeting can draw upon the thinking of many individuals, with the result being the solution of problems and near optimal decision making.

Often overlooked is the potential motivating force that meetings can have. There is much to indicate that meetings create a "oneness" and "psychic" benefit for the participants.

Coordination and Communication Meetings

In spite of the potential benefits of meetings, meetings tend to be too long, too much is attempted, and too limited planning is done. Too many meetings today are less than successful for one of two reasons.

The meeting should never have been held in the first place
The meeting was necessary but it was conducted improperly

It follows that a construction manager can put meetings to work by identifying and holding ones that are necessary and beneficial and taking steps to ensure that the meeting is conducted effectively and accomplishes its objectives.

Meetings can be classified into one of four types:

1. Report meeting
2. Decision-making meeting
3. Training meeting
4. Creative meeting

Each of these meetings requires varying degrees of leadership, freedom for the discussing members, and objectives. Several meetings of each of the four types are essential to the construction manager.

The report meeting should be guided by strong direction of the construction manager. It is designed for the direct presentation of material. The material presented could lead to a decision-making meeting either held spontaneously or at a later date. The report meeting should not be free-wheeling.

The number of report meetings that are part of the CM process is limited. Usually they are limited to meetings in which the CM firm updates its client on project progress. The construction manager might also find it advantageous to update contractors on the overall project progress via a report meeting.

The decision-making meeting is the most used and probably the most productive meeting held by the construction manager. It is specifically designed to draw together the thinking of the attendees into a decision. The most frequent and essential decision-making meetings are job-related meetings. These include *prejob meetings, job production meetings,* and *job safety meetings.*

The prejob meeting is usually both a report meeting and a decision-making meeting. It is a report meeting in that it reviews job conditions, contract clauses, manpower, and project layout with contractors and material suppliers. In addition to the information or reporting segment of the meeting, decisions regarding potential job problems are made. The key is to focus on potential problem areas to keep them from occurring so that they do not become part of a report meeting at a later date.

The job production meeting can take many different forms. Often it is a weekly meeting with project contractors to air problems and set out work assignments. The most effective time for the job production meeting is usually on Monday.

The job safety meeting has increased in importance with the increase in insurance costs and the existence of the Occupational Safety and Health Act. A safety meeting that addresses decisions designed to prevent accidents can have better returns than a safety meeting that merely exposes workers to the topic of safety. The job safety meeting can be effectively held in conjunction with job production meetings. Work being scheduled by the job production meeting should be addressed in the safety meeting.

Too often construction managers overlook the benefits of training and creative meetings. Training meetings have the objective of passing information in the simplest, most easily understood form. The training meeting can set a pattern for having contractors work together and develop an ease of talking together. Communication and participation of this type can play an important role in the personnel management program of the construction manager. The concept is to have contractors feel as if they are part of the team.

It may not be obvious why the construction manager would want to hold training meetings for its client's contractors. However, if the construction manager is to use management techniques and procedures effectively, it must have the cooperation and understanding of the contractors. For example, the construction manager benefits the project owner, the contractors, and itself, if it holds a CPM project planning and scheduling training/information meeting for the project contractors. The contractors should be shown the benefits of the CPM process, including overall project benefits as well as benefits that will flow to the contractors themselves. Only then will the construction manager obtain the cooperation of the contractors in regard to its planning and scheduling objectives.

Unlike the decision-making meeting, which is designed to resolve specific problems, the creative meeting attempts to develop new ideas and personal relationships advantageous to the continuity of a project. The creative meeting is sometimes referred to as *brainstorming*. The construction industry or construction manager has made little use of the creative meeting. No doubt this can be traced to the fact that it is usually difficult to measure the benefit from the creative meeting. Sometimes the benefits are immediate and apparent. More often, the benefits are received over a long period and difficult to measure.

The number of creative meetings compatible or useful in the CM process are limited. However, meetings between the construction manager and the design team that have the objective of critiquing the project contract documents might be viewed as creative meetings. It may also be advantageous for the construction manager to brainstorm with the project contractors with the objective of drawing out the expertise and skills of the contractors.

MONITORING PROJECT PERFORMANCE

The effective construction manager serves the project owner by being at the job site and exercising daily decision making over and with the contractor team. In effect it is in visual contact with the day-to-day work progress. However, visual observation is not all there is to the control function.

In addition to the construction manager's visual observation and project control, the firm must execute control through formal documentation of progress through the use of data control forms and procedures. Relevant to visual observation, these processed control forms have the following advantages

1. Provides the project owner with a communication link to project progress
2. Serves as a means of flagging potential problem areas and supplementing the visual observation of the construction manager
3. Provides a written document to serve as support for payments made contractors
4. Provides written evidence of interim contractor performance that can be used in any potential legal dispute regarding costs or time of performance

Written documentation of progress with an eye to controlling the project is one of the services that is set out in almost every owner/CM firm contract.

Unlike a contractor's control system that focuses primarily on monitoring costs, the construction manager's control reporting forms focus primarily on the documentation of the time of performance or work progress. The construction manager usually does not have access to an individual contractor's cost records. On the other hand, it can monitor the time of performance through its visual observation of work put in place. In effect, the construction manager's ability to control construction phase time and cost centers around its ability to keep a project on schedule.

The set of forms and procedures used by a construction manager in monitoring contractor performance is often referred to as a *cost control system* or *project information system* and the forms themselves as *control forms*. The processed forms serve as the means for the construction manager to compare planned progress with actual progress with the potential of identifying inefficiencies that need correction.

The first step in the successful monitoring of contractor performance is initiated with the preparation of a detailed project plan. The preparation of formalized plans using a bar chart and CPM were discussed in chapter 8. An example plan to serve the basis of a construction manager's control system is illustrated in Figure 10.1. Note that a plot of planned project owner cash flow is also illustrated. These cash plans can assist in obtaining optimal

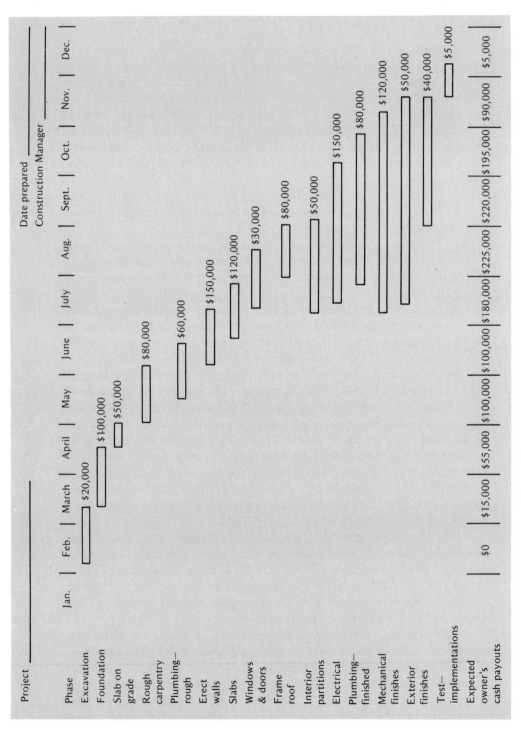

Figure 10:1 Planned progress—basis for control

Monitoring Project Performance

```
Project Name _____  Week Ending _____
                                                Name _____
I.   WORK PLANNED FOR WEEK (per CONSTRUCTION MANAGER PLAN)
                        Anticipated Work      Expected % Complete
         Phase          "This" Week           at End of Week
         1.
         2.
         3.
         4.

II.  ACTUAL WORK COMPLETED THIS WEEK
                        Quantity of Work      % Complete at
         Phase          "This" Week           End of Week
         1.
         2.
         3.
         4.

III. REASONS WHY SCHEDULE WAS NOT ATTAINED
         1.
         2.
         3.
         4.

IV.  INFORMATION RELEVANT TO NEXT WEEK'S WORK OR FUTURE
     WORK ON PROJECT
         1.
         2.
         3.
         4.
```

Figure 10:2 Weekly progress report for job site

financing for the project. Once a project is budgeted, a basis for control is provided. It is the benchmark by which actual performance is evaluated.

Actual performance is determined by direct observation of the construction manager or by means of forms filled out by individual contractors. An example of one of these forms is illustrated in Figure 10.2. The contractor also notes any difficulty or event worthy of the construction manager's attention on the form. The construction manager monitors the correctness of the progress reported on the forms through its day-to-day contact with the project.

The form shown in Figure 10.2 serves as the basis of the construction manager updating the progress to date via the example form illustrated in Figure 10.3. As shown, the form illustrates the actual progress versus the

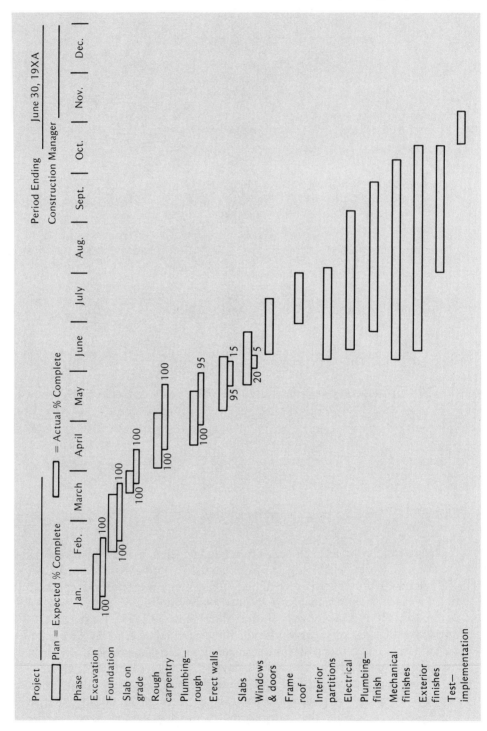

Figure 10:3 Planned versus actual progress

Monitoring Project Performance

planned progress. The form also accommodates the preparation of a revised plan of performance based on the report progress.

The setting out of planned progress on the example form shown in Figure 10.3 serves several purposes. For one it can signal performance or schedule problems early so that they can be corrected. A management technique for correcting them is illustrated in a later section of this chapter. The documentation of progress also serves as an aid to the construction manager in evaluating contractor payment requests. The construction manager's role in this function is discussed in the next section.

Project Name _____   Week Starting _____
                                             Contractor _____

I. WORK PLANNED FOR WEEK (LISTED IN ORDER OF PRIORITY)

Phase	Anticipated Work "This" Week	Expected % Complete at End of Week
1.		
2.		
3.		
4.		

II. EXPECTED RESOURCES REQUIRED TO ACHIEVE PLAN

Type	Amount
1.	
2.	
3.	
4.	

III. EXPECTED INTERFACE WITH OTHER CONTRACTORS

1.
2.

IV. POSSIBLE PROBLEM AREAS NEEDING ATTENTION

1.
2.
3.

Figure 10:4   Construction manager's plan of action for contractors

The construction manager can use the information in Figure 10.3 as the basis for determining its plan of action for the next time period of the project. This plan of action should include direction for the individual contractors. The example form shown in Figure 10.4 can be used to communicate this plan of action to the individual contractors. The process repeats itself weekly or monthly as the work progresses.

The forms illustrated in Figures 10.1–10.4 are only examples. The actual forms need to be custom designed to the individual CM firm. The CM firm's ability to collect data, process them, and use the information dictate the actual set of forms. However, the objective of the forms remains unchanged, to monitor progress with the objective of comparing it to planned progress with the intent of identifying or detecting potential problems.

The use of automated data processing equipment assists in the collection of more data and the quicker processing of them, but a computerized control system performs essentially the same calculation and purpose of a manual system. However, the computer has a larger memory capacity and can process data more quickly than a manual system. The same can be said of the construction manager's estimating function or planning function discussed in earlier chapters.

MONITORING CONTRACTOR PAYOUTS

The construction manager's role in supervising the construction phase of a project puts it in an excellent role to serve the owner in reviewing and approving contractor requests for payment. The construction manager's objectives in performing this function are to protect the owner's financial assets and to ensure that the contractors are justly and timely paid for work performed.

Without the CM process, approval of contractor payments is usually the responsibility of the architect/engineer design team. However, the design team's lack of daily involvement in a project's construction phase and the design team's close alignment with its client and somewhat adverse relationship with the contractor team restricts its objective and fair monitoring and approval of contractor payment requests. These conditions can lead to the design team and the owner falling victim to approval and payment of contractor overbillings. In other situations the design team may unjustly attempt to pressure contractors by holding back payment due. Although holding back contractor payments can be used as a means of pressuring a contractor to perform or serve to force an agreement on a dispute, such negative management is met by even more severe problems, including aggravating an already adverse relationship.

The fact that contracts are for work performed over a relatively long time period results in the need for contractor interim payment requests. For example, the general conditions of the contract documents often state that the contractor is permitted to make monthly progress billings for work performed and to receive payment for these requests.

The problem that can occur relevant to the contractor interim billings is whether the pay requests represent work performed or are in excess of fair compensation. The excess billings are referred to as an *overbillings* or a *front-*

ending of profit. The contractor's interest in front-ending its profit is to speed up the collection of cash that can be used to minimize finance costs or be invested in other operations.

The owner's concern in being overbilled is obvious. Any overbillings and subsequent payment are for work not performed. In effect the project owner is financing a contractor's operations. Secondly, given the fact that money earns income through interest, an owner's payment of an overbilling increases the owner's financing cost and therefore overall cost of the project.

Perhaps of more concern to a project owner and its lending institution is the potential legal exposure associated with contractor overpayments resulting from an overbillings. Should a contractor go bankrupt during performance of a construction project, the project owner and lending institution can take comfort in the fact that they have obtained the work paid for prior to the contractor's financial failure. However, if the payments include overbillings, the project owner and lending institution have no recourse to recapture the excess payment. In effect, the payment represents money that cannot be recovered in either cash or construction work.

As noted earlier, the construction manager's daily involvement in supervising a project's construction should aid it in monitoring and passing judgment on the fairness of contractor requested payments. Also, the fact that the CM process is characterized by the awarding of several separate construction contracts should enhance the evaluation of contractor payment requests. The breakout of construction packages allows for a more detailed determination of work complete relative to the total work contracted. In a general contractor process it is difficult to evaluate the fairness of progress payment requests made by the general contractor in behalf of its subcontractors.

The key elements to the determination of the fairness of a contractor pay request are the following:

1. Breakout of total work quantities for distinct items of the contracted work package
2. Breakout of contract dollar amounts per distinct item of the contracted work package
3. Determination of work performed to date of progress payment request per distinct item of the contracted work package
4. Necessary adjustments due to problem areas or unusual conditions

The breakout of a contractor's work package into distinct item quantities and dollar amounts provides the framework for an equitable determination of contractor earned income. For example, assume a contractor bids $500,000 to perform the concrete work for a specific project. Assume further that the project owner requests that the contractor breakout the bid as to work to perform as is illustrated in Figure 10.5.

Work Item	Quantity (cu. yds.)	$Bid
Footings	1,600	$ 80,000
Columns	1,000	70,000
Beams	1,250	50,000
Walls	1,000	100,000
Slabs	2,000	150,000
Sitework	2,000	50,000
		$500,000

Figure 10:5 Breakdown of work for concrete contract

Given the breakout shown, it becomes possible to evaluate contractor payment requests by a detailed analysis of the work performed. For example, assume that through July the work performed per item is as illustrated in Figure 10.6. Note that some of the items were determined on a quantity-complete basis, whereas others are on a percentage complete basis. Either will enable determination of an equitable contractor payment.

Work Item	Budget Quantity (cu. yd.)	Quantity or % Complete
Footings	1,600	100%
Columns	1,000	800 cu. yd.
Beams	1,250	50%
Walls	1,000	250 cu. yd.
Slabs	2,000	0%
Sitework	2,000	0%

Figure 10:6 Progress per construction manager review

Given the information set out in Figure 10.6, the calculation of payment due the concrete contractor for the work performed in July is as shown in Figure 10.7. This calculation is based on a work-item basis. The payment calculated can then be compared to the contractor's prepared payment request to evaluate it for fairness and the possibility of an overbilling.

As an alternative to the process illustrated in Figure 10.7, a contractor's payment request can be evaluated merely by passing judgment on the total costs incurred to date (assuming this information is available to the project

Monitoring Contractor Payouts

owner or construction manager) relevant to the total contract dollar amount. The error in this process relates to the fact that a contractor's interim project costs may not be proportional to the work put in place. For example, due to performance inefficiencies, a contractor could be incurring cost overruns. The result would be that a contractor payment based on costs incurred would in effect be an overpayment.

Work Item	Budget Quantity (cu. yd.)	Quantity or % Complete	$ Bid	$ Earned (due contractor)
Footings	1,600	100%	$ 80,000	$ 80,000
Columns	1,000	800 cu. yd.	70,000	56,000
Beams	1,250	50%	50,000	25,000
Walls	1,000	250 cu. yd.	100,000	25,000
Slabs	2,000	0%	150,000	0
Sitework	2,000	0%	50,000	0
			$500,000	$186,000

Assume: Owner-requested billing $200,000
         Amount due contractor   186,000
         Overbilling $ 14,000

Figure 10:7 Calculation of payment due contractor

Although the calculation of payment due the contractor illustrated in Figure 10.7 is straightforward, the construction manager has to be aware of conditions that can cause the calculation to be in error. For one, the contractor in setting out bid amounts for the items delineated in Figure 10.5 could attempt to weight the dollars of initial work with the intent of front-ending payment requests. For example, the dollar amount of the concrete slabs could be lessened in favor of increasing the dollar amount of the footing concrete, which is performed first. The solution is for the construction manager to evaluate the dollar amounts of the individual items by using cost-estimating skills at the time the payment schedule is prepared. Ability to do this is dependent on the detailedness of the item breakout.

The construction manager must be on the alert for conditions that can distort the determination of the work put in place as of a certain date. For example, the purchase of material that is stored at a project to be put up in place throughout the project's duration can complicate the process. For one, the determination of payment due the contractor can be too low if determined only by work actually put in place with no credit being given to stored materials. If the contractor is expected to purchase the material in bulk it has claim to payment for the cost of the material purchased independent of whether or not it is in place. On the other hand, a revision downward in regard to a contractor's payment request is needed if the contractor claims payment for both the material cost and its installation cost if in fact the material is stored at the job site. The end result is that the construction manager needs to be on the alert for conditions that result in a modification of the calculations illustrated in Figure 10.7.

CORRECTING INEFFICIENT CONSTRUCTION OPERATIONS

The fact that a construction manager detects a performance inefficiency does not in itself ensure its correction. The existence and noncorrection of the inefficiency can threaten the construction manager's ability to deliver the construction project to its client on time and within the budgeted cost.

The use of the control forms illustrated earlier should be viewed as the means of flagging nonperformance. By themselves they do not get at the cause of a construction operation inefficiency or suggest a possible cure. Without this identification the control function is incomplete.

The complexity of a construction operation necessitates a detailed analysis if one is to be able to identify the cause of an inefficiency or establish a possible cure. For example, the use of control forms might indicate that a prime contractor responsible for forming walls for concrete placement is falling behind schedule. The process of erecting wall forms includes the use of several carpenters, laborers, ironworkers, operating engineers, and equipment and material. Numerous problems could exist that might be corrected. For example, there could be a crew imbalance between the carpenters and laborers, the equipment could be inefficient, or perhaps unavailability of materials like form ties or formwork accessories is constraining productivity and causing delays. The identification of the actual cause is contingent upon an analysis of the process.

One might raise the question as to what benefit will result from the construction manager identifying inefficient construction processes or methods and further establishing a cure or means of improving the inefficiency. This question might be asked in light of the fact that a detected inefficiency is likely a contractor inefficiency and not the direct result of the construction manager. In effect, the construction manager could be unable to impose its recommendations on the contractor. This point was discussed earlier in the chapter.

Although there are constraints to the amount of pressure a construction manager can put on a contractor to increase productivity, the failure to perform this function places the construction manager in a position of being responsible to deliver construction phase time and cost with no means of attaining these objectives. The end result is that either through the authority it has as an agent of the project owner or through its ability to have the contractor team implement recommendations because it respects and wants to cooperate with the construction manager, the construction manager must

1. Detect possible inefficiencies
2. Determine the cause of the inefficiency
3. Identify a possible cure for the inefficiency
4. Use authority as CM or gain the cooperation of the contractors to have them implement changes for work improvement

The aggressive and effective construction manager will not wait until a problem occurs before it monitors and analyzes operations of the contractor team. On a periodic basis the construction manager will attempt to analyze construction operations with the objective of identifying means of work improvement that, if implemented, will aid the contractor in being profitable and will also result in the construction work being performed on schedule.

In effect, the construction manager is a troubleshooter for the project's contractors. If the construction manager can detect inefficiencies and determine recommendations for improvement, it can gain the respect of the contractor team. This respect is needed if the construction manager is to be effective in its control duties.

THE METHOD PRODUCTIVITY DELAY MODEL

The previous section discussed the construction manager's ability to analyze a construction operation or method as part of its control function. The question now becomes how does the construction manager perform this function.

Although its knowledge of construction is a valuable asset in critiquing a construction operation, the lack of a formalized approach can limit the benefits of analysis. Even more importantly, the lack of use of a formalized approach or management technique will result in the failure to even attempt to critique or analyze a construction operation or method.

Several management techniques have been established for assisting in the detailed analysis of a production or work process. Techniques like work sampling, time study, queuing models, motion analysis, and process charts are used by several manufacturing industries to study and analyze production. However, the fact that many of these focus only on labor, need to be applied at a very detailed time increment level, or are limited to the analysis of few work components that are part of the process, constrain or eliminate their use as techniques for analyzing a construction operation.

In this section a model will be discussed that has been developed by the author for modeling and improving construction productivity. The model is referred to as the Method Productivity Delay Model (MPDM). It can be viewed as an alternative to the more traditional work sampling, time study, and motion analysis models. The objective of the model is to provide the construction manager and the construction firm with a practical means of measuring, predicting, and improving productivity and hence the time it takes to complete a construction process.

An overview of the MPDM is shown in Figure 10.8. The model is broken into four elements, the collection of data, the processing of data, the structured model, and the implementation element. Each of these elements will now be discussed separately.

Collection of Data

The purpose of the collection element is to collect data to be used as a basis for modeling the method. It addresses the determination and collection of three types of information.

1. Identify the production unit
2. Identify the production cycle
3. Collect data concerning the time required for the completion of production cycles and document productivity delays

The definition of a method's production unit serves as the basis for measuring, predicting, and improving the method. The *production unit* is an amount of work descriptive of the production that can easily be measured. Typical examples of production units are

1. Arrival of a scraper in a borrow-pit
2. Release of concrete from a crane bucket
3. Placement of one row of concrete blocks on a wall
4. Placement of a structural member

A *production cycle* is the time between consecutive occurrences of the production unit. The production unit cannot be defined independent of the other elements of the model. The definition of the production unit dictates the detail used to measure the method productivity. If the production unit is defined in too broad a context, the collected productiviy information is of little value in construction method measurement, prediction, and improvement. For example, if the production unit is defined as the completion of the placing and finishing of concrete for a wall section of 200 square feet, the collected production cycle times and delay information can be too broad to focus on variables that affect productivity or inefficiencies. On the other hand, too detailed a definition of the production unit can curtail the construction manager's ability to identify productivity delays or even the completion of a production cycle. For example, if the placing of a single brick is identified as the production unit when four masons are simultaneously placing bricks,

The Method Productivity Delay Model

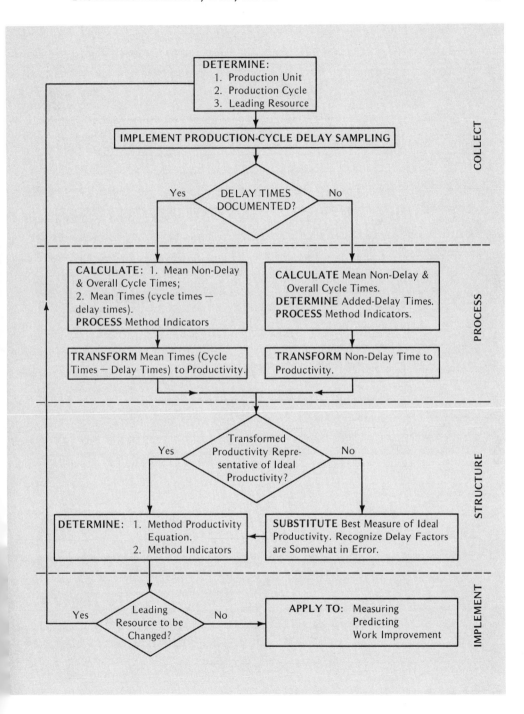

Figure 10:8 Overview of the MPDM model

PRODUCTION CYCLE DELAY SAMPLING

Page 1 of 1
Method: CRANE BUCKET CONCRETE POUR
Unit: Second
Production Unit: Concrete Drop

Production Cycle	Production Cycle Time (sec.)	Environment Delay	Equipment Delay	Labor Delay	Material Delay	Management Delay	Notes	Minus Mean Non-delay Time
1	120			✓				27
2	126		✓					33
3	98						✓	5
4	112		✓					19
5	108						✓	15
6	1122					✓	crane move	1029
7	116		✓					23
8	214		✓					121
9	92						✓	1
10	88						✓	5
11	100						✓	7
12	312		✓					219
13	110			✓				17
14	666		90%	10%				573
15	146		25%	75%				53
16	120		✓					27
17	138		✓					45
18	144		20%	80%				51
19	598	✓					crane slip	505
20	118		✓					25
21	138		✓					45
22	108						✓	15
23	98						✓	5
24	120				✓			27
25	116				✓			23
26	368		✓					275
27	118		✓					25
28	140		✓					47
29	136				✓			43
30	138				✓			45

Figure 10:9 Data collection using the MPDM

The Method Productivity Delay Model

31	154	✓			61
32	396	✓			303
33	96			✓	3
34	286	✓			193
35	80			✓	13
36	82			✓	11
37	84			✓	9
38	212		✓		119
39	78			✓	15
40					

Figure 10:9 *continued*

the model user will find it difficult to document the successive completion of production cycles, let alone cite productivity delays.

The data collected in documenting the production cycle times and delays serve as the basis for the delay information part of the productivity equation that is part of the MPDM. The documented production cycle times serve as a basis for the first part of the productivity equation.

The following are documented:

1. Time required to complete production cycle
2. Occurrence of productivity delay
3. If more than one productivity delay type is found in a given cycle, the total delay of the cycle is allocated by approximate percentage or by documented times
4. Any unusual events that characterize a given production cycle

A data collection form along with example data is shown in Figure 10.9. The procedure for collecting the data is referred to as *production cycle delay sampling* (PCDS).

Documenting production cycle times is straightforward. The construction manager merely clocks the time between occurrences of the production unit. If the method is very complex or the production cycle times are very short, such that visual inspection is impossible or inaccurate, filming procedures such as time lapse can be used to document the method productivity.

Citing and documenting production cycle delays, whether they are single or multiple delays, requires a degree of skill from the MPDM user. The construction manager's ability to single out productivity delay types normally increases with experience and decreases as the complexity of the method in question increases. The types of delays identified vary depending on the construction method. Construction method delays can normally be identified that are observable and independent of one another. The following example of the MPDM will consider five delay types: *environment, equipment, labor, material,* and *management*. If the construction manager finds it convenient to add or delete from these five types, the MPDM procedure remains unchanged. Examples of the given types of delays follow.

1. Environment: Change in soil conditions, change in wall section, change in roadway alignment
2. Equipment: Stationary production equipment in transit, equipment operating at less than capacity
3. Labor: Workmen waiting for other workmen, workmen loafing, worker fatigue, workmen not productive because of lack of knowledge of work
4. Material: Material not available for equipment or labor demand, material defective
5. Management: Poor planning of method resource combination and placement, secondary operation interfering with method productivity, poor method layout planning

Model Processing and Structuring

The processing element of the MPDM provides the connection among the model's collected data, the structured method productivity equation, and the accompanying measure of risk and variability. The processing consists of adding, subtracting, multiplying, and dividing; the procedure possesses the desired model attributes of ease and economy of implementation. The processing data formulated from the collected data shown in Figure 10.9 are processed by means of the form shown in Figure 10.10. The form and entries are self-explanatory.

The first part of the structured MPDM as shown in Figure 10.11 is a method productivity equation that relates overall or actual method productivity to ideal method productivity as a function of the identified productivity delay types. This method productivity equation is used directly to aid in measuring, predicting, and improving method productivity.

Ideal productivity is that productivity occurring when productivity delays are absent. The nondelay production cycles documented in the collection of data (Figure 10.9) are, in fact, ideal productivity cycles if all delays have been cited. However, if conditions exist that indicate some delays are not being detected, the construction manager should select other techniques for determining ideal productivity. Even if the nondelay cycles appear adequate as a measure of ideal productivity, it may be beneficial to compare the calculated productivity with a historical record.

MPDM PROCESSING

Method: Crane Bucket Concrete Pour
Production Unit: Concrete Drop
Unit: second

	Units	Total Production Time	Number of Cycles	Mean Cycle Time	$\Sigma[|(\text{Cycle time}) - (\text{Nondelay cycle time})|]/n$
A.	Nondelayed production cycles	1112	12	92.6	8.7
B.	Overall production cycles	7596	39	194.7	104.5

Delay Information

		Environment	Equipment	Labor	Material	Management
C.	Occurrences	1	17	11	0	1
D.	Total added time	505	1939	500	0	1029
E.	Probability of occurrences[a]	.026	.436	.282	0	.026
F.	Relative severity[b]	2.59	.58	.23	0	5.28
G.	Expected % delay time per production cycle[c]	6.7	25.4	6.6	0	13.7

[a] Delay cycles/total number of cycles
[b] Mean added cycle time/mean overall cycle time
[c] Row E times row F times 100

Figure 10:10 Processing of data using the MPDM

The nondelay production cycle time is transformed into production per time period for which productivity is to be measured, predicted, and work improvement is to be performed. This time period will normally be in units per hour or day. The mean nondelay production cycle time for the example is 92.6 seconds. The ideal productivity calculated on an hourly basis is

$$\frac{(60 \text{ min/hr}) \times (60 \text{ sec/min})}{92.6 \text{ sec/unit}} = 38.9 \text{ units/hr}$$

MPDM STRUCTURE

Crane Bucket Concrete Pour

Production Unit: Concrete Drop

I. Productivity Equation
Overall method
productivity = (Ideal productivity) $(1 - E_{en} - E_{eq} - E_{la} - E_{mt} - E_{mn})$
18.5 units/hr = (38.9 units/hr) $(1 - .067 - .254 - .066 - 0.137)$

where: E_{en} = expected environmental delay time as a decimal fraction of total production time

E_{eq} = expected equipment delay time

E_{la} = expected labor delay time

E_{mt} = expected material delay time

E_{mn} = expected management delay time

II. Method Indicators

A. Variability of method productivity
Ideal cycle variability = 8.7/92.6 = 0.09
Overall cycle variability = 104.5/194.7 = 0.54

B. Delay information

	Environment	Equipment	Labor	Material	Management
Probability of occurrence	0.026	0.436	0.287	0	0.026
Relative severity	2.59	0.58	0.23	0	5.28
Expected % delay time per production cycle	6.7	25.4	6.6	0	13.7

Figure 10:11 Structured data using the MPDM

If the calculated value is not considered reliable, the value determined in some other manner can be used as the ideal productivity. If the production unit is appropriately defined and the construction manager is skillful in detecting errors, this substitution will seldom be warranted.

The environmental (E_{en}), equipment (E_{eq}), labor (E_{la}), material (E_{mt}), and management (E_{mn}) factors in the right-hand side of the productivity equation shown in Figure 10.11 relate the method's ideal productivity to the method's overall productivity. The factors, which can have values ranging from 0 to 1, are the decimal fraction of total production time caused by each delay type. The factor 1 minus the sum of E_{en}, E_{eq}, E_{la}, E_{mt}, and E_{mn} relates the probability of productivity work being performed.

The Method Productivity Delay Model

The left-hand side of the productivity equation in Figure 10.11 is the *overall method productivity*.

Part 2 of the structured model shown in Figure 10.11 contains *method indicators*. The information set out in this segment will be used as an indicator to predict and improve method productivity.

Four types of information are set out in the *method indicator* part of the structured MPDM. The first, *variability of method productivity*, gives a measure of the variable nature of both the nondelay productivity cycles and the total overall productivity cycles. These are determined as follows.

$$\text{Ideal Cycle Variability} = \frac{(\text{Nondelay cycle time}) - (\text{Mean nondelay cycle time})}{\text{Number of nondelay cycles}} \bigg/ \text{Mean nondelay cycle time}$$

$$\text{Overall Cycle Variability} = \frac{(\text{Overall cycle time}) - (\text{Mean nondelay cycle time})}{\text{Number of total cycles}} \bigg/ \text{Mean overall cycle time}$$

Nondelay cycle times are the durations of cycles in which no delays are detected. Overall cycle times are the durations of all cycles without regard to detection of delays. For the data shown in Figure 10.9 and processed in Figure 10.10, the variability of method productivity is 0.09 for the ideal cycles and 0.54 for overall cycles.

The three other indicators in this part of the model relate to selected productivity delays. They are the *probability of occurrence, relative severity,* and the *expected percent of delay time per production cycle*. These indicators are used in analysis of potential work improvement.

Model Implementation and Benefits

The calculated MPDM nondelay and overall productivity rates represent measured productivities. Measuring is one of the objectives of the model. The derived productivity equation provides the basis for method productivity prediction. The factors relating method ideal productivity to overall method productivity can be predicted or calculated from past data.

The constant in the method productivity equation, the ideal productivity, can be considered fixed. For a given construction method some variation exists in the time of ideal production cycles; however, this variability is small compared to the variability in the cycle times for overall production cycles.

The delay factors can all be considered variables. Although the delay factors (Es) are given values as a result of collection of data from the previous performance(s) of the method in question, they can take on new values for a future occurrence of the construction method.

When a construction manager is analyzing a construction method for which MPDM data has been collected, attention should be focused on the calculated delay factors. He or she might determine that one or more of the factors will be lower or higher on the upcoming performance of the method. The previously determined delay factors should therefore be adjusted. This adjustment will result in prediction of a new overall method productivity.

The construction manager cannot assume that actual results will correspond to predicted values. If the method's leading resource is changed in quantity, a new method results and MPDM prediction becomes difficult. If only support resources are changed, prediction of productivity is enhanced. For example, three support laborers might be used on a construction method and the MPDM for the method might indicate a 30 percent labor delay factor. How much of a reduction in the delay factor will occur if an additional laborer is used for the method? The MPDM does not yield a single prediction; however, it does indicate that the method productivity should increase between 0 and 30 percent. Further consideration of the interdependencies among the method's resource variables and the other delay factors should provide the construction manager with a means of estimating the actual reduction in the delay factor and the corresponding increase in method productivity.

Once the construction productivity for the new method is documented, the prediction can be evaluated. This evaluation should serve as a means of better predicting future method productivity.

Of the three tasks of measuring, predicting, and improving method productivity, the MPDM is probably the most useful to the construction manager in the work-improvement task. Consider the structured MPDM in Figure 10.11, which was derived from the data collected in Figure 10.9 and processed in Figure 10.10. Inspection of the MPDM structure for the method indicates that the equipment delay and the management delay are critical as they result in large percentage production delay times. Attention should be focused on these two types of delays when trying to improve productivity.

The MPDM provides the construction manager the capability to set out construction operation or method inefficiencies in a formal process. By means of quantifying the amount of non-productive time and the inefficiencies of a process, the construction manager has a means of communicating this to the contractor. The fact that the construction manager criticizes a contractor's production process will not mean that the contractor will attempt

Exercise

to correct the situation. In fact, the criticism may be viewed negatively. However, should the criticism be accompanied by a detailed but simple quantification of results, as can be accomplished with the MPDM, the possibility that the construction manager will get the attention and cooperation of the contractor is increased.

The MPDM should not be viewed as an optimal model; however, it is a formal, quantified process. Such a process can assist the contractors in helping them attain their profit objectives and thus help the construction manager and its client. It is in this manner that the aggressive CM firm applies effective construction management.

EXERCISE 10.1

MONITORING CONTRACTOR PAYMENTS

Data regarding the budget and to-date performance through June 30, 19XA, for various work phases or packages for a project are shown in the accompanying table. Calculate for each work phase or package 1) the amount of revenue earned through June by the contractor; 2) any over- or underbillings through June; 3) expected cash to be paid by the owner for each work phase or package in July, assuming all worked not billed through June will be completed and billed in July and that all overbillings through June will be paid in July; and 4) the percentage complete for the total project through June based on the ratio of earned contractor revenue through June divided by the total contract dollar amount. Assume there is no retainer held on work performed and the contractor is paid for the work in the same month it performs the work.

Work Package or Phase	Dollar Contract Amount	Total Contractor Billings through May	Contractor Billing in June	Total Cash Paid to Contractor through May	% Complete CM Firm's Determination through June
Concrete footings	$ 200,000	$100,000	$100,000	$ 90,000	100%
Concrete walls	400,000	150,000	100,000	150,000	50%
Concrete slabs	500,000	20,000	80,000	15,000	25%
Concrete beams	100,000	10,000	40,000	10,000	40%
	$1,200,000	$280,000	$320,000	$265,000	

EXERCISE 10.2

CORRECTING A JOB SITE INEFFICIENCY

Assume that a construction manager collects the data in Figure 10.12 regarding the performance of one of its project contractor's construction methods. The construction manager has the objective of making recommendations to the contractor to enhance the contractor's productivity and also to ensure that the contractor keeps on schedule. Process the data in the figure to determine the nonproductive time by cause, and the overall hourly productivity of the contractor.

Production Cycle	Production Cycle Time	Material	Labor	Equipment	Management	Ideal	Delay
		PRODUCTION CYCLE DELAY SAMPLING					
		Method: Pour Concrete: 1 unit = 1 cu. yd. Production Unit: Concrete Bucket Drop					
1	98					✓	
2	78					✓	
3	212	✓					
4	96					✓	
5	120		✓				
6	176	✓					
7	190			✓			
8	88					✓	
9	184				✓		
10	78					✓	
11	124		✓				
12	136		✓				
13							
14							
15							
16							
17							
18							
19							
20							
21							
22							
23							
24							
25							
26							
27							
28							
29							
30							
31							
32							
33							

Figure 10:12

APPENDIX A

AIA Standard CM Contract

THE AMERICAN INSTITUTE OF ARCHITECTS

AIA Document B801

Standard Form of Agreement Between Owner and Construction Manager

Recommended for use with the current editions of standard AIA agreement forms and documents.

THIS DOCUMENT HAS IMPORTANT LEGAL CONSEQUENCES; CONSULTATION WITH AN ATTORNEY IS ENCOURAGED WITH RESPECT TO ITS COMPLETION OR MODIFICATION

AGREEMENT

made this   day of   in the year of Nineteen Hundred and

BETWEEN the Owner:

and the Construction Manager:

For services in connection with the following described Project:
(Include detailed description of Project location and scope)

The Architect for the Project is:

The Owner and the Construction Manager agree as set forth below.

AIA DOCUMENT B801 • OWNER-CONSTRUCTION MANAGER AGREEMENT • DECEMBER 1973 EDITION • AIA®
© 1973 • THE AMERICAN INSTITUTE OF ARCHITECTS, 1735 NEW YORK AVE., N.W., WASHINGTON, D. C. 20006

Appendix A

I. THE CONSTRUCTION MANAGER, as agent of the Owner, shall provide services in accordance with the Terms and Conditions of this Agreement.

The Construction Manager accepts the relationship of trust and confidence established between him and the Owner by this Agreement. He covenants with the Owner to furnish his professional skill and judgment and to cooperate with the Architect in furthering the interests of the Owner. He agrees to furnish efficient business administration and superintendence and to use his professional efforts at all times in an expeditious and economical manner consistent with the interests of the Owner.

II. THE OWNER shall compensate the Construction Manager in accordance with the Terms and Conditions of this Agreement.

    A. *FOR BASIC SERVICES*, as described in Paragraph 1.1, compensation shall be computed as follows: **(select one)**

        1. On the basis of a PROFESSIONAL FEE PLUS EXPENSES.

           a. A Professional Fee of

                                                                                               dollars ($),

           PLUS

           b. An amount computed as follows:

              Principals' time at the fixed rate of

                                                                              dollars ($) per hour.

              For the purposes of this Agreement, the Principals are:

              Employees' time (other than Principals) at the following multiples of the employees' Direct Personnel Expense as defined in Article 4.

                  For those assigned to the Project in the office, a multiple of

                                                                                  (   ).

                  For those assigned to the Project at the construction site, a multiple of

                                                                                    (   ).

           c. Services of professional consultants at a multiple of

                                                                            (   ) times

              the amount billed to the Construction Manager for such services.

AIA DOCUMENT B801 • OWNER-CONSTRUCTION MANAGER AGREEMENT • DECEMBER 1973 EDITION • AIA®
© 1973 • THE AMERICAN INSTITUTE OF ARCHITECTS, 1735 NEW YORK AVE., N.W., WASHINGTON, D. C. 20006

AIA Standard CM Contract

OR 2. On the basis of a MULTIPLE OF DIRECT PERSONNEL EXPENSE.

    a. Principals' time at the fixed rate of
                        dollars ($) per hour.
       For the purposes of this Agreement, the Principals are:

    b. Employees' time (other than Principals) at the following multiples of the employees' Direct Personnel Expense as defined in Article 4.

       For those assigned to the Project in the office, a multiple of (   ).

       For those assigned to the Project at the construction site, a multiple of (   ).

    c. Services of professional consultants at a multiple of (   ) times the amount billed to the Construction Manager for such services.

OR 3. On the basis of a FIXED FEE of
                        dollars ($).

OR 4. On the basis of a PERCENTAGE OF CONSTRUCTION COST, as defined in Article 3, of
                        percent (%).

  B. *AN INITIAL PAYMENT* of
                        dollars ($).
shall be made upon the execution of this Agreement and credited to the Owner's account.

  C. *FOR ADDITIONAL SERVICES*, as described in Paragraph 1.2, compensation shall be computed as follows:

    1. Principals' time at the fixed rate of
                        dollars ($) per hour.
       For the purposes of this Agreement, the Principals are:

    2. Employees' time (other than Principals) at the following multiples of the employees' Direct Personnel Expense as defined in Article 4.

       For those assigned to the Project in the office, a multiple of (   ).

       For those assigned to the Project at the construction site, a multiple of (   ).

    3. Services of professional consultants at a multiple of (   ) times the amount billed to the Construction Manager for such services.

Appendix A

D. *FOR REIMBURSABLE EXPENSES,* amounts expended as defined in Article 5.

III. THE PARTIES agree in accordance with the Terms and Conditions of this Agreement that:

A. *IF THE SCOPE* of the Project is changed materially, compensation for Basic Services shall be subject to renegotiation.

B. *IF THE SERVICES* covered by this Agreement have not been completed within ( ) months of the date hereof, the amounts of compensation, rates and multiples set forth in Paragraph II shall be subject to renegotiation.

AIA Standard CM Contract

> TERMS AND CONDITIONS OF AGREEMENT BETWEEN
> OWNER AND CONSTRUCTION MANAGER

ARTICLE 1

CONSTRUCTION MANAGER'S SERVICES

1.1 BASIC SERVICES

The Construction Manager's Basic Services consist of the two phases described below and any other services included in Article 13 as Basic Services.

DESIGN PHASE

1.1.1 *Consultation During Project Development:* Review conceptual designs during development. Advise on site use and improvements, selection of materials, building systems and equipment. Provide recommendations on relative construction feasibility, availability of materials and labor, time requirements for installation and construction, and factors related to cost including costs of alternative designs or materials, preliminary budgets, and possible economies.

1.1.2 *Scheduling:* Provide and periodically update a Project time schedule that coordinates and integrates the Architect's services with construction schedules.

1.1.3 *Project Budget:* Prepare a Project budget for the Owner's approval as soon as major Project requirements have been identified and update periodically. Prepare an estimate of construction cost based on a quantity survey of Drawings and Specifications at the end of the Schematic Design Phase for approval by the Owner. Update and refine this estimate for Owner's approval as the development of the Drawings and Specifications proceeds, and advise the Owner and the Architect if it appears that the Project budget will not be met and make recommendations for corrective action.

1.1.4 *Coordination of Contract Documents:* Review the Drawings and Specifications as they are being prepared, recommending alternative solutions whenever design details affect construction feasibility or schedules.

1.1.4.1 Verify that the requirements and assignment of responsibilities for safety precautions and programs, temporary Project facilities and for equipment, materials and services for common use of Contractors are included in the proposed Contract Documents.

1.1.4.2 Advise on the method to be used for selecting Contractors and awarding contracts. If separate contracts are to be awarded, review the Drawings and Specifications to (1) ascertain if areas of jurisdiction overlap, (2) verify that all Work has been included, and (3) allow for phased construction.

1.1.4.3 Investigate and recommend a schedule for purchase by the Owner of all materials and equipment requiring long lead time procurement, and coordinate the schedule with the early preparation of Contract Documents by the Architect. Expedite and coordinate delivery of these purchases.

1.1.5 *Labor:* Provide an analysis of the types and quantity of labor required for the Project and review the availability of appropriate categories of labor required for critical phases.

1.1.5.1 Determine applicable requirements for equal employment opportunity programs for inclusion in the proposed Contract Documents.

1.1.6 *Bidding:* Prepare pre-qualification criteria for bidders and develop Contractor interest in the Project. Establish bidding schedules and conduct pre-bid conferences to familiarize bidders with the bidding documents and management techniques and with any special systems, materials or methods.

1.1.6.1 Receive bids, prepare bid analyses and make recommendations to the Owner for award of contracts or rejection of bids.

1.1.7 *Contract Awards:* Conduct pre-award conferences with successful bidders. Assist the Owner in preparing Construction Contracts and advise the Owner on the acceptability of Subcontractors and material suppliers proposed by Contractors.

CONSTRUCTION PHASE

The Construction Phase will commence with the award of the first Construction Contract or purchase order and will terminate 30 days after the final Certificate for Payment is issued by the Architect.

1.1.8 *Project Control:* Coordinate the Work of the Contractors with the activities and responsibilities of the Owner and Architect to complete the Project in accordance with the Owner's objectives on cost, time and quality. Provide sufficient personnel at the Project site with authority to achieve these objectives.

1.1.8.1 Schedule and conduct pre-construction and progress meetings at which Contractors, Owner, Architect and Construction Manager can discuss jointly such matters as procedures, progress, problems and scheduling.

1.1.8.2 Provide a detailed schedule for the operations of Contractors on the Project, including realistic activity sequences and durations, allocation of labor and materials, processing of shop drawings and samples, and delivery of products requiring long lead time procurement; include the Owner's occupancy requirements showing portions of the Project having occupancy priority.

1.1.8.3 Provide regular monitoring of the schedule as construction progresses. Identify potential variances between scheduled and probable completion dates. Review schedule for Work not started or incomplete and recommend to the Owner and Contractor adjustments in the schedule to meet the probable completion date. Provide summary reports of each monitoring, and document all changes in schedule.

1.1.8.4 Recommend courses of action to the Owner when requirements of a contract are not being fulfilled.

AIA DOCUMENT B801 • OWNER-CONSTRUCTION MANAGER AGREEMENT • DECEMBER 1973 EDITION • AIA®
© 1973 • THE AMERICAN INSTITUTE OF ARCHITECTS, 1735 NEW YORK AVE., N.W., WASHINGTON, D. C. 20006

1.1.9 *Cost Control:* Revise and refine the approved estimate of construction cost, incorporate approved changes as they occur, and develop cash flow reports and forecasts as needed.

1.1.9.1 Provide regular monitoring of the approved estimate of construction cost, showing actual costs for activities in process and estimates for uncompleted tasks. Identify variances between actual and budgeted or estimated costs, and advise the Owner and Architect whenever projected costs exceed budgets or estimates.

1.1.9.2 Arrange for the maintenance of cost accounting records on authorized Work performed under unit costs, actual costs for labor and materials, or other bases requiring accounting records.

1.1.9.3 Develop and implement a system for review and processing of Change Orders.

1.1.9.4 Recommend necessary or desirable changes to the Owner and the Architect, review requests for changes, submit recommendations to the Owner and the Architect, and assist in negotiating Change Orders.

1.1.9.5 Develop and implement a procedure for the review and processing of applications by Contractors for progress and final payments. Make recommendations to the Architect for certification to the Owner for payment.

1.1.10 *Permits and Fees:* Assist in obtaining all building permits and special permits for permanent improvements, excluding permits for inspection or temporary facilities required to be obtained directly by the various Contractors. Verify that the Owner has paid all applicable fees and assessments for permanent facilities. Assist in obtaining approvals from all the authorities having jurisdiction.

1.1.11 *Owner's Consultants:* If required, assist the Owner in selecting and retaining professional services of a surveyor, special consultants and testing laboratories. Coordinate these services.

1.1.12 *Inspection:* Inspect the work of Contractors to assure that the Work is being performed in accordance with the requirements of the Contract Documents. Endeavor to guard the Owner against defects and deficiencies in the Work. Require any Contractor to stop Work or any portion thereof, and require special inspection or testing of any Work not in accordance with the provisions of the Contract Documents whether or not such Work be then fabricated, installed or completed. Reject Work which does not conform to the requirements of the Contract Documents.

1.1.12.1 The Construction Manager shall not be responsible for construction means, methods, techniques, sequences and procedures employed by Contractors in performance of their contract, and he shall not be responsible for the failure of any Contractors to carry out the Work in accordance with the Contract Documents.

1.1.13 *Contract Performance:* Consult with the Architect and the Owner if any Contractor requests interpretations of the meaning and intent of the Drawings and Specifications, and assist in the resolution of any questions which may arise.

1.1.14 *Shop Drawings and Samples:* In collaboration with the Architect, establish and implement procedures for expediting the processing and approval of shop drawings and samples.

1.1.15 *Reports and Records:* Record the progress of the Project. Submit written progress reports to the Owner and the Architect including information on the Contractors and Work, the percentage of completion and the number and amounts of Change Orders. Keep a daily log available to the Owner and the Architect.

1.1.15.1 Maintain at the Project site, on a current basis: records of all Contracts; shop drawings; samples; purchases; materials; equipment; applicable handbooks; federal, commercial and technical standards and specifications; maintenance and operating manuals and instructions; and any other related documents and revisions which arise out of the Contract or the Work. Obtain data from Contractors and maintain a current set of record drawings, specifications and operating manuals. At the completion of the Project, deliver all such records to the Owner.

1.1.16 *Owner-Purchased Items:* Accept delivery and arrange storage, protection and security for all Owner-purchased materials, systems and equipment which are a part of the Work until such items are turned over to the Contractors.

1.1.17 *Substantial Completion:* Upon the Contractors' determination of Substantial Completion of the Work or designated portions thereof, prepare for the Architect a list of incomplete or unsatisfactory items and a schedule for their completion. After the Architect certifies the Date of Substantial Completion, supervise the correction and completion of Work.

1.1.18 *Start-Up:* With the Owner's maintenance personnel, direct the checkout of utilities, operational systems and equipment for readiness and assist in their initial start-up and testing.

1.1.19 *Final Completion:* Determine final completion and provide written notice to the Owner and Architect that the Work is ready for final inspection. Secure and transmit to the Architect required guarantees, affidavits, releases, bonds and waivers. Turn over to the Owner all keys, manuals, record drawings and maintenance stocks.

1.2 ADDITIONAL SERVICES

The following Additional Services shall be performed upon authorization in writing from the Owner and shall be paid for as hereinbefore provided.

1.2.1 Services related to investigation, appraisals or valuations of existing conditions, facilities or equipment, or verifying the accuracy of existing drawings or other Owner-furnished information.

1.2.2 Services related to Owner-furnished equipment, furniture and furnishings which are not a part of the Work.

1.2.3 Services for tenant or rental spaces.

1.2.4 Services related to construction performed by the Owner.

1.2.5 Consultation on replacement of Work damaged by

AIA Standard CM Contract

fire or other cause during construction, and furnishing services for the replacement of such Work.

1.2.6 Services made necessary by the default of a Contractor.

1.2.7 Preparing to serve or serving as an expert witness in connection with any public hearing, arbitration proceeding, or legal proceeding.

1.2.8 Finding housing for construction labor, and defining requirements for establishment and maintenance of base camps.

1.2.9 Obtaining or training maintenance personnel or negotiating maintenance service contracts.

1.2.10 Services related to Work items required by the Conditions of the Contract and the Specifications which are not provided by Contractors.

1.2.11 Inspections of and services related to the Project after completion of the services under this Agreement.

1.2.12 Providing any other service not otherwise included in this Agreement.

ARTICLE 2

THE OWNER'S RESPONSIBILITIES

2.1 The Owner shall provide full information regarding his requirements for the Project.

2.2 The Owner shall designate a representative who shall be fully acquainted with the scope of the Work, and has authority to render decisions promptly and furnish information expeditiously.

2.3 The Owner shall retain an Architect for design and to prepare construction documents for the Project. The Architect's services, duties and responsibilities are described in the Agreement between the Owner and the Architect, pertinent parts of which will be furnished to the Construction Manager and will not be modified without written notification to him.

2.4 The Owner shall furnish such legal, accounting and insurance counselling services as may be necessary for the Project, and such auditing services as he may require to ascertain how or for what purposes the Contractors have used the moneys paid to them under the Construction Contracts.

2.5 The Owner shall furnish the Construction Manager with a sufficient quantity of construction documents.

2.6 If the Owner becomes aware of any fault or defect in the Project or nonconformance with the Contract Documents, he shall give prompt written notice thereof to the Construction Manager.

2.7 The services, information, surveys and reports required by Paragraphs 2.3 through 2.5 inclusive shall be furnished at the Owner's expense, and the Construction Manager shall be entitled to rely upon the accuracy and completeness thereof.

ARTICLE 3

CONSTRUCTION COST

3.1 If the Construction Cost is to be used as the basis for determining the Construction Manager's compensation for Basic Services, it shall be the cost of all Work, including work items in the Conditions of the Contract and the Specifications and shall be determined as follows:

3.1.1 For completed construction, the total construction cost of all such Work;

3.1.2 For Work not constructed, (1) the sum of the lowest bona fide bids received from qualified bidders for any or all of such Work or (2) if the Work is not bid, the sum of the bona fide negotiated proposals submitted for any or all of such Work; or

3.1.3 For Work for which no such bids or proposals are received, the Construction Cost contained in the Construction Manager's latest Project construction budget approved by the Owner.

3.2 Construction Cost shall not include the compensation of the Construction Manager (except for costs of Work items in the Conditions of the Contract and in the Specifications), the Architect and consultants, the cost of the land, rights-of-way, or other costs which are the responsibility of the Owner as provided in Paragraphs 2.3 through 2.5 inclusive.

3.3 The cost of labor, materials and equipment furnished by the Owner shall be included in the Construction Cost at current market rates, including a reasonable allowance for overhead and profit.

3.4 Cost estimates prepared by the Construction Manager represent his best judgment as a professional familiar with the construction industry. It is recognized, however, that neither the Construction Manager nor the Owner has any control over the cost of labor, materials or equipment, over Contractors' methods of determining bid prices, or other competitive bidding or market conditions.

3.5 No fixed limit of Construction Cost shall be deemed to have been established unless it is in writing and signed by the parties hereto. When a fixed limit of Construction Cost is established in writing as a condition of this Agreement, the Construction Manager shall advise what materials, equipment, component systems and types of construction should be included in the Contract Documents, and shall suggest reasonable adjustments in the scope of the Project to bring it within the fixed limit.

3.5.1. If responsive bids are not received as scheduled, any fixed limit of Construction Cost established as a condition of this Agreement shall be adjusted to reflect any change in the general level of prices occurring between the originally scheduled date and the date on which bids are received.

3.5.2 When the fixed limit of Construction Cost is exceeded, the Owner shall (1) give written approval of an increase in such fixed limit, (2) authorize rebidding within a reasonable time, or (3) cooperate in revising the scope and quality of the Work as required to reduce the Construction Cost. In the case of (3) the Construction Manager, without additional compensation, shall cooperate with the Architect as necessary to bring the Construction Cost within the fixed limit.

ARTICLE 4

DIRECT PERSONNEL EXPENSE

Direct Personnel Expense is defined as the salaries of professional, technical and clerical employees engaged on the Project by the Construction Manager, and the cost of their mandatory and customary benefits such as statutory employee benefits, insurance, sick leave, holidays, vacations, pensions and similar benefits.

ARTICLE 5

REIMBURSABLE EXPENSES

5.1 Reimbursable Expenses are in addition to the compensation for Basic and Additional Services and include actual expenditures made by the Construction Manager, his employees, or his professional consultants in the interest of the Project:

5.1.1 Long distance calls and telegrams, and fees paid for securing approval of authorities having jurisdiction over the Project.

5.1.2 Handling, shipping, mailing and reproduction of Project related materials.

5.1.3 Transportation and living when traveling in connection with the Project, relocation costs, and overtime work requiring higher than regular rates if authorized in advance by the Owner.

5.1.4 Electronic data processing service and rental of electronic data processing equipment when used in connection with Additional Services.

5.1.5 Premiums for insurance required in Article 10.

5.1.6 Providing construction support activities such as Work items included in the Conditions of the Contract and in the Specifications unless they are provided by the Contractors.

ARTICLE 6

PAYMENTS TO THE CONSTRUCTION MANAGER

6.1 Payments shall be made monthly upon presentation of the Construction Manager's statement of services as follows:

6.1.1 An initial payment as set forth in Paragraph IIB is the minimum payment under this Agreement.

6.1.2 When compensation is computed as described in Paragraphs IIA1a, IIA3, or IIA4, subsequent payments for Basic Services shall be made in proportion to services performed. The compensation at the completion of each Phase shall equal the following percentages of the total compensation for Basic Services:

Design Phase	20%
Construction Phase	100%

6.1.3 Payments for Reimbursable Expenses shall be made upon presentation of the Construction Manager's statement.

6.2 No deductions shall be made from the Construction Manager's compensation on account of penalty, liquidated damages, or other sums withheld from payments to Contractors.

6.3 If the Project is suspended for more than three months or abandoned in whole or in part, the Construction Manager shall be paid his compensation for services performed prior to receipt of written notice from the Owner of such suspension or abandonment, together with Reimbursable Expenses then due, and all termination expenses as defined in Paragraph 7.2 resulting from such suspension or abandonment. If the Project is resumed after being suspended for more than three months, the Construction Manager's compensation shall be subject to renegotiation.

6.4 If construction of the Project has started and is delayed by reason of strikes, or other circumstance not due to the fault of the Construction Manager, the Owner shall reimburse the Construction Manager for the costs of his Project-site staff as provided for by this Agreement. The Construction Manager shall reduce the size of his Project-site staff after a 60-day delay, or sooner if feasible, for the remainder of the delay period as directed by the Owner and, during the period, the Owner shall reimburse the Construction Manager for the direct personnel expense of such staff plus any relocation costs. Upon the termination of the delay, the Construction Manager shall restore his Project-site staff to its former size, subject to the approval of the Owner.

6.5 If the Project time schedule established in Subparagraph 1.1.2 is exceeded by more than thirty days through no fault of the Construction Manager, compensation for Basic Services performed by Principals, employees and professional consultants required beyond the thirtieth day to complete the services under this Agreement shall be as set forth in Paragraph IIC.

6.6 Payments due the Construction Manager which are unpaid for more than 60 days from date of billing shall bear interest at the legal rate of interest applicable at the construction site.

ARTICLE 7

TERMINATION OF AGREEMENT

7.1 This Agreement may be terminated by either party upon seven days' written notice should the other party fail substantially to perform in accordance with its terms through no fault of the party initiating termination. In the event of termination due to the fault of others than the Construction Manager, the Construction Manager shall be paid his compensation plus Reimbursable Expenses for services performed to termination date and all termination expenses.

7.2 Termination expenses are defined as Reimbursable Expenses directly attributable to termination, plus an

AIA Standard CM Contract

amount computed as a percentage of the total compensation earned to the time of termination, as follows:

- 20 percent if termination occurs during the Design Phase; or
- 10 percent if termination occurs during the Construction Phase.

ARTICLE 8

SUCCESSORS AND ASSIGNS

The Owner and the Construction Manager each binds himself, his partners, successors, assigns and legal representatives to the other party to this Agreement and to the partners, successors, assigns and legal representatives of such other party with respect to all covenants of this Agreement. Neither the Owner nor the Construction Manager shall assign, sublet or transfer his interest in this Agreement without the written consent of the other.

ARTICLE 9

ARBITRATION

9.1 All claims, disputes and other matters in question between the parties to this Agreement, arising out of, or relating to this Agreement or the breach thereof, shall be decided by arbitration in accordance with the Construction Industry Arbitration Rules of the American Arbitration Association then obtaining unless the parties mutually agree otherwise. No arbitration, arising out of, or relating to this Agreement, shall include, by consolidation, joinder or in any other manner, any additional party not a party to this Agreement except by written consent containing a specific reference to this Agreement and signed by all the parties hereto. Any consent to arbitration involving an additional party or parties shall not constitute consent to arbitration of any dispute not described therein or with any party not named or described therein. This Agreement to arbitrate and any agreement to arbitrate with an additional party or parties duly consented to by the parties hereto shall be specifically enforceable under the prevailing arbitration law.

9.2 Notice of the demand for arbitration shall be filed in writing with the other party to this Agreement and with the American Arbitration Association. The demand shall be made within a reasonable time after the claim, dispute or other matter in question has arisen. In no event shall the demand for arbitration be made after the date when institution of legal or equitable proceedings based on such claim, dispute or other matter in question would be barred by the applicable statute of limitations.

9.3 The award rendered by the arbitrators shall be final, and judgment may be entered upon it in accordance with applicable law in any court having jurisdiction thereof.

ARTICLE 10

INSURANCE

The Construction Manager shall purchase and maintain insurance to protect himself from claims under workmen's compensation acts; claims for damages because of bodily injury including personal injury, sickness or disease, or death of any of his employees or of any person other than his employees; and from claims for damages because of injury to or destruction of tangible property including loss of use resulting therefrom; and from claims arising out of the performance of professional services caused by any errors, omissions or negligent acts for which he is legally liable.

ARTICLE 11

EXTENT OF AGREEMENT

11.1 This Agreement represents the entire and integrated agreement between the Owner and the Construction Manager and supersedes all prior negotiations, representations or agreements, either written or oral. This Agreement shall not be superseded by provisions of contracts for construction and may be amended only by written instrument signed by both Owner and Construction Manager.

11.2 Nothing contained herein shall be deemed to create any contractual relationship between the Construction Manager and the Architect or any of the Contractors, Subcontractors or material suppliers on the Project; nor shall anything contained herein be deemed to give any third party any claim or right of action against the Owner or the Construction Manager which does not otherwise exist without regard to this Agreement.

ARTICLE 12

GOVERNING LAW

Unless otherwise specified, this Agreement shall be governed by the law in effect at the location of the Project.

Appendix A

ARTICLE 13

OTHER CONDITIONS OR SERVICES

This Agreement executed the day and year first written above.

OWNER   CONSTRUCTION MANAGER

_____   _____

AIA DOCUMENT B801 • OWNER-CONSTRUCTION MANAGER AGREEMENT • DECEMBER 1973 EDITION • AIA®
© 1973 • THE AMERICAN INSTITUTE OF ARCHITECTS, 1735 NEW YORK AVE., N.W., WASHINGTON, D. C. 20006

APPENDIX B

AGC Standard CM Contract

THE ASSOCIATED GENERAL CONTRACTORS

STANDARD FORM OF AGREEMENT BETWEEN OWNER AND CONSTRUCTION MANAGER

(GUARANTEED MAXIMUM PRICE OPTION)

(See AGC Document No. 8a for Fixing the
Guaranteed Maximum Price and
AGC Document 8b for recommended General Conditions)

This Document has important legal and insurance consequences; consultation with an attorney is encouraged with respect to its completion or modification.

AGREEMENT

Made this   day of   in the year of Nineteen Hundred and

BETWEEN

the Owner, and

the Construction Manager.

For services in connection with the following described Project: (Include complete Project location and scope)

The Architect/Engineer for the Project is

The Owner and the Construction Manager agree as set forth below:

Certain provisions of this document have been derived, with modifications, from the following documents published by The Amerian Institute of Architects: AIA Document A111, Owner-Contractor Agreement, © 1974; AIA Document A201, General Conditions, © 1976; AIA Document B801, Owner-Construction Manager Agreement, © 1973, by The American Institute of Architects. Usage made of AIA language, with the permission of AIA, does not apply AIA endorsement or approval of this document. Further reproduction of copyrighted AIA materials without separate written permission from AIA is prohibited.

AGC Document No. 8   Owner-Construction Manager Agreement   June 1977
© 1977 Associated General Contractors of America

ARTICLE 1

The Construction Team and Extent of Agreement

The CONSTRUCTION MANAGER accepts the relationship of trust and confidence established between him and the Owner by this Agreement. He covenants with the Owner to furnish his best skill and judgment and to cooperate with the Architect/Engineer in furthering the interests of the Owner. He agrees to furnish efficient business administration and superintendence and to use his best efforts to complete the Project in the best and soundest way and in the most expeditious and economical manner consistent with the interest of the Owner.

1.1 *The Construction Team:* The Construction Manager, the Owner, and the Architect/Engineer called the "Construction Team" shall work from the beginning of design through construction completion. The Construction Manager shall provide leadership to the Construction Team on all matters relating to construction.

1.2 *Extent of Agreement:* This Agreement represents the entire agreement between the Owner and the Construction Manager and supersedes all prior negotiations, representations or agreements. When Drawings and Specifications are complete, they shall be identified by amendment to this Agreement. This Agreement shall not be superseded by any provisions of the documents for construction and may be amended only by written instrument signed by both Owner and Construction Manager.

1.3 *Definitions:* The Project is the total construction to be performed under this Agreement. The Work is that part of the construction that the Construction Manager is to perform with his own forces or that part of the construction that a particular Trade Contractor is to perform. The term day shall mean calendar day unless otherwise specifically designated.

ARTICLE 2

Construction Manager's Services

The Construction Manager will perform the following services under this Agreement in each of the two phases described below.

2.1 Design Phase

2.1.1 *Consultation During Project Development:* Schedule and attend regular meetings with the Architect/Engineer during the development of conceptual and preliminary design to advise on site use and improvements, selection of materials, building systems and equipment. Provide recommendations on construction feasibility, availability of materials and labor, time requirements for installation and construction, and factors related to cost including costs of alternative designs or materials, preliminary budgets, and possible economies.

2.1.2 *Scheduling:* Develop a Project Time Schedule that coordinates and integrates the Architect/Engineer's design efforts with construction schedules. Update the Project Time Schedule incorporating a detailed schedule for the construction operations of the Project, including realistic activity sequences and durations, allocation of labor and materials, processing of shop drawings and samples, and delivery of products requiring long lead-time procurement. Include the Owner's occupancy requirements showing portions of the Project having occupancy priority.

2.1.3 *Project Construction Budget:* Prepare a Project budget as soon as major Project requirements have been identified, and update periodically for the Owner's approval. Prepare an estimate based on a quantity survey of Drawings and Specifications at the end of the schematic design phase for approval by the Owner as the Project Construction Budget. Update and refine this estimate for Owner's approval as the development of the Drawings and Specifications proceeds, and advise the Owner and the Architect/Engineer if it appears that the Project Construction Budget will not be met and make recommendations for corrective action.

2.1.4 *Coordination of Contract Documents:* Review the Drawings and Specifications as they are being prepared, recom-

AGC Standard CM Contract

mending alternative solutions whenever design details affect construction feasibility or schedules without, however, assuming any of the Architect/Engineer's responsibilities for design.

2.1.5 *Construction Planning:* Recommend for purchase and expedite the procurement of long-lead items to ensure their delivery by the required dates.

2.1.5.1 Make recommendations to the Owner and the Architect/Engineer regarding the division of Work in the Drawings and Specifications to facilitate the bidding and awarding of Trade Contracts, allowing for phased construction taking into consideration such factors as time of performance, availability of labor, overlapping trade jurisdictions, and provisions for temporary facilities.

2.1.5.2 Review the Drawings and Specifications with the Architect/Engineer to eliminate areas of conflict and overlapping in the Work to be performed by the various Trade Contractors and prepare prequalification criteria for bidders.

2.1.5.3 Develop Trade Contractor interest in the Project and as working Drawings and Specifications are completed, take competitive bids on the Work of the various Trade Contractors. After analyzing the bids, either award contracts or recommend to the Owner that such contracts be awarded.

2.1.6 *Equal Employment Opportunity:* Determine applicable requirements for equal employment opportunity programs for inclusion in Project bidding documents.

2.2 Construction Phase

2.2.1 *Project Control:* Monitor the Work of the Trade Contractors and coordinate the Work with the activities and responsibilities of the Owner, Architect/Engineer and Construction Manager to complete the Project in accordance with the Owner's objectives of cost, time and quality.

2.2.1.1 Maintain a competent full-time staff at the Project site to coordinate and provide general direction of the Work and progress of the Trade Contractors on the Project.

2.2.1.2 Establish on-site organization and lines of authority in order to carry out the overall plans of the Construction Team.

2.2.1.3 Establish procedures for coordination among the Owner, Architect/Engineer, Trade Contractors and Construction Manager with respect to all aspects of the Project and implement such procedures.

2.2.1.4 Schedule and conduct progress meetings at which Trade Contractors, Owner, Architect/Engineer and Construction Manager can discuss jointly such matters as procedures, progress, problems and scheduling.

2.2.1.5 Provide regular monitoring of the schedule as construction progresses. Identify potential variances between scheduled and probable completion dates. Review schedule for Work not started or incomplete and recommend to the Owner and Trade Contractors adjustments in the schedule to meet the probable completion date. Provide summary reports of each monitoring and document all changes in schedule.

2.2.1.6 Determine the adequacy of the Trade Contractors' personnel and equipment and the availability of materials and supplies to meet the schedule. Recommend courses of action to the Owner when requirements of a Trade Contract are not being met.

2.2.2 *Physical Construction:* Provide all supervision, labor, materials, construction equipment, tools and subcontract items which are necessary for the completion of the Project which are not provided by either the Trade Contractors or the Owner. To the extent that the Construction Manager performs any Work, with his own forces, he shall, with respect to such Work, be bound to the extent not inconsistent with this Agreement, by the procedures and the obligations with respect to such Work as may govern the Trade Contractors under any General Conditions to the Trade Contracts.

2.2.3 *Cost Control:* Develop and monitor an effective system of Project cost control. Revise and refine the initially approved Project Construction Budget, incorporate approved changes as they occur, and develop cash flow reports and forecasts as needed. Identify variances between actual and budgeted or estimated costs and advise Owner and Architect/Engineer whenever projected cost exceeds budgets or estimates.

2.2.3.1 Maintain cost accounting records on authorized Work performed under unit costs, actual costs for labor and material, or other bases requiring accounting records. Afford the Owner access to these records and preserve them for a period of three (3) years after final payment.

2.2.4 *Change Orders:* Develop and implement a system for the preparation, review and processing of Change Orders. Recommend necessary or desirable changes to the Owner and the Architect/Engineer, review requests for changes, submit recommendations to the Owner and the Architect/Engineer, and assist in negotiating Change Orders.

2.2.5 *Payments to Trade Contractors:* Develop and implement a procedure for the review, processing and payment of applications by Trade Contractors for progress and final payments.

2.2.6 *Permits and Fees:* Assist the Owner and Architect/Engineer in obtaining all building permits and special permits for permanent improvements, excluding permits for inspection or temporary facilities required to be obtained directly by the various Trade Contractors. Assist in obtaining approvals from all the authorities having jurisdiction.

2.2.7 *Owner's Consultants:* If required, assist the Owner in selecting and retaining professional services of a surveyor, testing laboratories and special consultants, and coordinate these services.

2.2.8 *Inspection:* Inspect the Work of Trade Contractors for defects and deficiencies in the Work without assuming any of the Architect/Engineer's responsibilities for inspection.

2.2.8.1 Review the safety programs of each of the Trade Contractors and make appropriate recommendations. In making such recommendations and carrying out such reviews, he shall not be required to make exhaustive or continuous inspections to check safety precautions and programs in connection with the Project. The performance of such services by the Construction Manager shall not relieve the Trade Contractors of their responsibilities for the safety of persons and property, and for compliance with all federal, state and local statutes, rules, regulations and orders applicable to the conduct of the Work.

2.2.9 *Document Interpretation:* Refer all questions for interpretation of the documents prepared by the Architect/Engineer to the Architect/Engineer.

2.2.10 *Shop Drawings and Samples:* In collaboration with the Architect/Engineer, establish and implement procedures for expediting the processing and approval of shop drawings and samples.

2.2.11 *Reports and Project Site Documents:* Record the progress of the Project. Submit written progress reports to the Owner and the Architect/Engineer including information on the Trade Contractors' Work, and the percentage of completion. Keep a daily log available to the Owner and the Architect/Engineer.

2.2.11.1 Maintain at the Project site, on a current basis: records of all necessary Contracts, Drawings, samples, purchases, materials, equipment, maintenance and operating manuals and instructions, and other construction related documents, including all revisions. Obtain data from Trade Contractors and maintain a current set of record Drawings, Specifications and operating manuals. At the completion of the Project, deliver all such records to the Owner.

2.2.12 *Substantial Completion:* Determine Substantial Completion of the Work or designated portions thereof and prepare for the Architect/Engineer a list of incomplete or unsatisfactory items and a schedule for their completion.

2.2.13 *Start-Up:* With the Owner's maintenance personnel, direct the checkout of utilities, operations systems and equipment for readiness and assist in their initial start-up and testing by the Trade Contractors.

2.2.14 *Final Completion:* Determine final completion and provide written notice to the Owner and Architect/Engineer that

AGC Standard CM Contract

the Work is ready for final inspection. Secure and transmit to the Architect/Engineer required guarantees, affidavits, releases, bonds and waivers. Turn over to the Owner all keys, manuals, record drawings and maintenance stocks.

2.2.15 *Warranty:* Where any Work is performed by the Construction Manager's own forces or by Trade Contractors under contract with the Construction Manager, the Construction Manager shall warrant that all materials and equipment included in such Work will be new, unless otherwise specified, and that such Work will be of good quality, free from improper workmanship and defective materials and in conformance with the Drawings and Specifications. With respect to the same Work, the Construction Manager further agrees to correct all work defective in material and workmanship for a period of one year from the Date of Substantial Completion or for such longer periods of time as may be set forth with respect to specific warranties contained in the trade sections of the Specifications. The Construction Manager shall collect and deliver to the Owner any specific written warranties given by others.

2.3 Additional Services

2.3.1 At the request of the Owner the Construction Manager will provide the following additional services upon written agreement between the Owner and Construction Manager defining the extent of such additional services and the amount and manner in which the Construction Manager will be compensated for such additional services.

2.3.2 Services related to investigation, appraisals or valuations of existing conditions, facilities or equipment, or verifying the accuracy of existing drawings or other Owner-furnished information.

2.3.3 Services related to Owner-furnished equipment, furniture and furnishings which are not a part of this Agreement.

2.3.4 Services for tenant or rental spaces not a part of this Agreement.

2.3.5 Obtaining or training maintenance personnel or negotiating maintenance service contracts.

ARTICLE 3

Owner's Responsibilities

3.1 The Owner shall provide full information regarding his requirements for the project.

3.2 The Owner shall designate a representative who shall be fully acquainted with the project and has authority to approve Project Construction Budgets, Changes in the Project, render decisions promptly and furnish information expeditiously.

3.3 The Owner shall retain an Architect/Engineer for design and to prepare construction documents for the Project. The Architect/Engineer's services, duties and responsibilities are described in the Agreement between the Owner and the Architect/Engineer, a copy of which will be furnished to the Construction Manager. The Agreement between the Owner and the Architect/Engineer shall not be modified without written notification to the Construction Manager.

3.4 The Owner shall furnish for the site of the Project all necessary surveys describing the physical characteristics, soil reports and subsurface investigations, legal limitations, utility locations, and a legal description.

3.5 The Owner shall secure and pay for necessary approvals, easements, assessments and charges required for the construction, use or occupancy of permanent structures or for permanent changes in existing facilities.

3.6 The Owner shall furnish such legal services as may be necessary for providing the items set forth in Paragraph 3.5, and such auditing services as he may require.

3.7 The Construction Manager will be furnished without charge all copies of Drawings and Specifications reasonably necessary for the execution of the Work.

3.8 The Owner shall provide the insurance for the Project as provided in Paragraph 12.4, and shall bear the cost of any bonds required.

3.9 The services, information, surveys and reports required by the above paragraphs shall be furnished with reasonable promptness at the Owner's expense, and the Construction Manager shall be entitled to rely upon the accuracy and completeness thereof.

3.10 If the Owner becomes aware of any fault or defect in the Project or non-conformance with the Drawings and Specifications, he shall give prompt written notice thereof to the Construction Manager.

3.11 The Owner shall furnish reasonable evidence satisfactory to the Construction Manager that sufficient funds are available and committed for the entire cost of the Project. Unless such reasonable evidence is furnished, the Construction Manager is not required to commence any Work, or may, if such evidence is not presented within a reasonable time, stop the Project upon 15 days notice to the Owner.

3.12 The Owner shall communicate with the Trade Contractors only through the Construction Manager.

ARTICLE 4

Trade Contracts

4.1 All portions of the Project that the Construction Manager does not perform with his own forces shall be performed under Trade Contracts. The Construction Manager shall request and receive proposals from Trade Contractors and Trade Contracts will be awarded after the proposals are reviewed by the Architect/Engineer, Construction Manager and Owner.

4.2 If the Owner refuses to accept a Trade Contractor recommended by the Construction Manager, the Construction Manager shall recommend an acceptable substitute and the Guaranteed Maximum Price if applicable shall be increased or decreased by the difference in cost occasioned by such substitution and an appropriate Change Order shall be issued.

4.3 Unless otherwise directed by the Owner, Trade Contracts will be between the Construction Manager and the Trade Contractors. Whether the Trade Contracts are with the Construction Manager or the Owner, the form of the Trade Contracts including the General and Supplementary Conditions shall be satisfactory to the Construction Manager.

4.4 The Construction Manager shall be responsible to the Owner for the acts and omissions of his agents and employees, Trade Contractors performing Work under a contract with the Construction Manager, and such Trade Contractors' agents and employees.

ARTICLE 5

Schedule

5.1 The services to be provided under this Contract shall be in general accordance with the following schedule:

5.2 At the time a Guaranteed Maximum Price is established, as provided for in Article 6, a Date of Substantial Completion of the project shall also be established.

5.3 The Date of Substantial Completion of the Project or a designated portion thereof is the date when construction is sufficiently complete in accordance with the Drawings and Specifications so the Owner can occupy or utilize the Project or designated portion thereof for the use for which it is intended. Warranties called for by this Agreement or by the Drawings and Specifications shall commence on the Date of Substantial Completion of the Project or designated portion thereof.

AGC Standard CM Contract

5.4 If the Construction Manager is delayed at any time in the progress of the Project by any act or neglect of the Owner or the Architect/Engineer or by any employee of either, or by any separate contractor employed by the Owner, or by changes ordered in the Project, or by labor disputes, fire, unusual delay in transportation, adverse weather conditions not reasonably anticipatable, unavoidable casualties or any causes beyond the Construction Manager's control, or by delay authorized by the Owner pending arbitration, the Construction Completion Date shall be extended by Change Order for a reasonable length of time.

ARTICLE 6

Guaranteed Maximum Price

6.1 When the design, Drawings and Specifications are sufficiently complete, the Construction Manager will, if desired by the Owner, establish a Guaranteed Maximum Price, guaranteeing the maximum price to the Owner for the Cost of the Project and the Construction Manager's Fee. Such Guaranteed Maximum Price will be subject to modification for Changes in the Project as provided in Article 9 and for additional costs arising from delays caused by the Owner or the Architect/Engineer.

6.2 When the Construction Manager provides a Guaranteed Maximum Price, the Trade Contracts will either be with the Construction Manager or will contain the necessary provisions to allow the Construction Manager to control the performance of the Work.

6.3 The Guaranteed Maximum Price will only include those taxes in the Cost of the Project which are legally enacted at the time the Guaranteed Maximum Price is established.

ARTICLE 7

Construction Manager's Fee

7.1 In consideration of the performance of the Contract, the Owner agrees to pay the Construction Manager in current funds as compensation for his services a Construction Manager's Fee as set forth in Subparagraphs 7.1.1 and 7.1.2.

7.1.1 For the performance of the Design Phase services, a fee of   which shall be paid monthly, in equal proportions, based on the scheduled Design Phase time.

7.1.2 For work or services performed during the Construction Phase, a fee of   which shall be paid proportionately to the ratio the monthly payment for the Cost of the Project bears to the estimated cost. Any balance of this fee shall be paid at the time of final payment.

7.2 Adjustments in Fee shall be made as follows:

7.2.1 For Changes in the Project as provided in Article 9, the Construction Manager's Fee shall be adjusted as follows:

7.2.2 For delays in the Project not the responsibility of the Construction Manager, there will be an equitable adjustment in the fee to compensate the Construction Manager for his increased expenses.

7.2.3 The Construction Manager shall be paid an additional fee in the same proportion as set forth in 7.2.1 if the Construction Manager is placed in charge of the reconstruction of any insured or uninsured loss.

7.3 Included in the Construction Manager's Fee are the following:

7.3.1 Salaries or other compensation of the Construction Manager's employees at the principal office and branch offices, except employees listed in Subparagraph 8.2.2.

7.3.2 General operating expenses of the Construction Manager's principal and branch offices other than the field office.

7.3.3 Any part of the Construction Manager's capital expenses, including interest on the Construction Manager's capital employed for the project.

7.3.4 Overhead or general expenses of any kind, except as may be expressly included in Article 8.

7.3.5 Costs in excess of the Guaranteed Maximum Price.

ARTICLE 8

Cost of the Project

8.1 The term Cost of the Project shall mean costs necessarily incurred in the Project during either the Design or Construction Phase, and paid by the Construction Manager, or by the Owner if the Owner is directly paying Trade Contractors upon the Construction Manager's approval and direction. Such costs shall include the items set forth below in this Article.

8.1.1 The Owner agrees to pay the Construction Manager for the Cost of the Project as defined in Article 8. Such payment shall be in addition to the Construction Manager's Fee stipulated in Article 7.

8.2 Cost Items

8.2.1 Wages paid for labor in the direct employ of the Construction Manager in the performance of his Work under applicable collective bargaining agreements, or under a salary or wage schedule agreed upon by the Owner and Construction Manager, and including such welfare or other benefits, if any, as may be payable with respect thereto.

8.2.2 Salaries of the Construction Manager's employees when stationed at the field office, in whatever capacity employed, employees engaged on the road in expediting the production or transportation of materials and equipment, and employees in the main or branch office performing the functions listed below:

8.2.3 Cost of all employee benefits and taxes for such items as unemployment compensation and social security, insofar as such cost is based on wages, salaries, or other remuneration paid to employees of the Construction Manager and included in the Cost of the Project under Subparagraphs 8.2.1 and 8.2.2.

8.2.4 The proportion of reasonable transportation, traveling, moving, and hotel expenses of the Construction Manager or of his officers or employees incurred in discharge of duties connected with the Project.

8.2.5 Cost of all materials, supplies and equipment incorporated in the Project, including costs of transportation and storage thereof.

8.2.6 Payments made by the Construction Manager or Owner to Trade Contractors for their Work performed pursuant to contract under this Agreement.

8.2.7 Cost, including transportation and maintenance, of all materials, supplies, equipment, temporary facilities and hand tools not owned by the workmen, which are employed or consumed in the performance of the Work, and cost less salvage value on such items used but not consumed which remain the property of the Construction Manager.

8.2.8 Rental charges of all necessary machinery and equipment, exclusive of hand tools, used at the site of the Project, whether rented from the Construction Manager or other, including installation, repairs and replacements, dismantling, removal, costs of lubrication, transportation and delivery costs thereof, at rental charges consistent with those prevailing in the area.

8.2.9 Cost of the premiums for all insurance which the Construction Manager is required to procure by this Agreement or is deemed necessary by the Construction Manager.

AGC Standard CM Contract

8.2.10 Sales, use, gross receipts or similar taxes related to the Project imposed by any governmental authority, and for which the Construction Manager is liable.

8.2.11 Permit fees, licenses, tests, royalties, damages for infringement of patents and costs of defending suits therefor, and deposits lost for causes other than the Construction Manager's negligence. If royalties or losses and damages, including costs of defense, are incurred which arise from a particular design, process, or the product of a particular manufacturer or manufacturers specified by the Owner or Architect/Engineer, and the Construction Manager has no reason to believe there will be infringement of patent rights, such royalties, losses and damages shall be paid by the Owner and not considered as within the Guaranteed Maximum Price.

8.2.12 Losses, expenses or damages to the extent not compensated by insurance or otherwise (including settlement made with the written approval of the Owner).

8.2.13 The cost of corrective work subject, however, to the Guaranteed Maximum Price.

8.2.14 Minor expenses such as telegrams, long-distance telephone calls, telephone service at the site, expressage, and similar petty cash items in connection with the Project.

8.2.15 Cost of removal of all debris.

8.2.16 Cost incurred due to an emergency affecting the safety of persons and property.

8.2.17 Cost of data processing services required in the performance of the services outlined in Article 2.

8.2.18 Legal costs reasonably and properly resulting from prosecution of the Project for the Owner.

8.2.19 All costs directly incurred in the performance of the Project and not included in the Construction Manager's Fee as set forth in Paragraph 7.3.

ARTICLE 9

Changes in the Project

9.1 The Owner, without invalidating this Agreement, may order Changes in the Project within the general scope of this Agreement consisting of additions, deletions or other revisions, the Guaranteed Maximum Price, if established, the Construction Manager's Fee and the Construction Completion Date being adjusted accordingly. All such Changes in the project shall be authorized by Change Order.

9.1.1 A Change Order is a written order to the Construction Manager signed by the Owner or his authorized agent issued after the execution of this Agreement, authorizing a Change in the Project and/or an adjustment in the Guaranteed Maximum Price, the Construction Manager's Fee, or the Construction Completion Date. Each adjustment in the Guaranteed Maximum Price resulting from a Change Order shall clearly separate the amount attributable to the Cost of the Project and the Construction Manager's Fee.

9.1.2 The increase or decrease in the Guaranteed Maximum Price resulting from a Change in the Project shall be determined in one or more of the following ways:

.1 by mutual acceptance of a lump sum properly itemized and supported by sufficient substantiating data to permit evaluation;

.2 by unit prices stated in the Agreement or subsequently agreed upon;

.3 by cost as defined in Article 8 and a mutually acceptable fixed or percentage fee; or

.4 by the method provided in Subparagraph 9.1.3.

9.1.3 If none of the methods set forth in Clauses 9.1.2.1 through 9.1.2.3 is agreed upon, the Construction Manager, provided he receives a written order signed by the Owner, shall promptly proceed with the Work involved. The cost of such Work shall then be determined on the basis of the reasonable expenditures and savings of those performing the Work attributed to the change, including, in the case of an increase in the Guaranteed Maximum Price, a reasonable increase in the Construction Manager's Fee. In such case, and also under Clauses 9.1.2.3 and 9.1.2.4 above, the Construction Manager shall keep and present, in such form as the Owner may prescribe, an itemized accounting together with appropriate supporting data of the increase in the Cost of the Project as outlined in Article 8. The amount of decrease in the Guaranteed Maximum Price to be allowed by the Construction Manager to the Owner for any deletion or change which results in a net decrease in cost will be the amount of the actual net decrease. When both additions and credits are invovled in any one change, the increase in Fee shall be figured on the basis of net increase, if any.

9.1.4 If unit prices are stated in the Agreement or subsequently agreed upon, and if the quantities originally contemplated are so changed in a proposed Change Order that application of the agreed unit prices to the quantities of Work proposed will cause substantial inequity to the Owner or the Construction Manager, the applicable unit prices and Guaranteed Maximum Price shall be equitably adjusted.

9.1.5 Should concealed conditions encountered in the performance of the Work below the surface of the ground or should concealed or unknown conditions in an existing structure be at variance with the conditions indicated by the Drawings, Specifications, or Owner-furnished information or should unknown physical conditions below the surface of the ground or should concealed or unknown conditions in an existing structure of an unusual nature, differing materially from those ordinarily encountered and generally recognized as inherent in work of the character provided for in this Agreement, be encountered, the Guaranteed Maximum Price and the Construction Completion Date shall be equitably adjusted by Change Order upon claim by either party made within a reasonable time after the first observance of the conditions.

9.2 Claims for Additional Cost or Time

9.2.1 If the Construction Manager wishes to make a claim for an increase in the Guaranteed Maximum Price, an increase in his fee, or an extension in the Construction Completion Date, he shall give the Owner written notice thereof within a reasonable time after the occurrence of the event giving rise to such claim. This notice shall be given by the Construction Manager before proceeding to execute any Work, except in an emergency endangering life or property in which case the Construction Manager shall act, at his discretion, to prevent threatened damage, injury or loss. Claims arising from delay shall be made within a reasonable time after the delay. No such claim shall be valid unless so made. If the Owner and the Construction Manager cannot agree on the amount of the adjustment in the Guaranteed Maximum Price, Construction Manager's Fee or Construction Completion Date, it shall be determined pursuant to the provisions of Article 16. Any change in the Guaranteed Maximum Price, Construction Manager's Fee or Construction Completion Date resulting from such claim shall be authorized by Change Order.

9.3 Minor Changes in the Project

9.3.1 The Architect/Engineer will have authority to order minor Changes in the Project not involving an adjustment in the Guaranteed Maximum Price or an extension of the Construction Completion Date and not inconsistent with the intent of the Drawings and Specifications. Such Changes may be effected by written order and shall be binding on the Owner and the Construction Manager.

9.4 Emergencies

9.4.1 In any emergency affecting the safety of persons or property, the Construction Manager shall act, at his discretion, to prevent threatened damage, injury or loss. Any increase in the Guaranteed Maximum Price or extension of time claimed by the Construction Manager on account of emergency work shall be determined as provided in this Article.

ARTICLE 10

Discounts

All discounts for prompt payment shall accrue to the Owner to the extent the Cost of the project is paid directly by the

AGC Standard CM Contract

Owner or from a fund made available by the Owner to the Construction Manager for such payments. To the extent the Cost of the Project is paid with funds of the Construction Manager, all cash discounts shall accrue to the Construction Manager. All trade discounts, rebates and refunds, and all returns from sale of surplus materials and equipment, shall be credited to the Cost of the Project.

ARTICLE 11

Payments to the Construction Manager

11.1 The Construction Manager shall submit monthly to the Owner a statement, sworn to if required, showing in detail all moneys paid out, costs accumulated or costs incurred on account of the Cost of the Project during the previous month and the amount of the Construction Manager's Fee due as provided in Article 7. Payment by the Owner to the Construction Manager of the statement amount shall be made within ten (10) days after it is submitted.

11.2 Final payment constituting the unpaid balance of the Cost of the Project and the Construction Manager's Fee shall be due and payable when the Project is delivered to the Owner, ready for beneficial occupancy, or when the Owner occupies the Project, whichever event first occurs, provided that the Project be then substantially completed and this Agreement substantially performed. If there should remain minor items to be completed, the Construction Manager and Architect/Engineer shall list such items and the Construction Manager shall deliver, in writing, his unconditional promise to complete said items within a reasonable time thereafter. The Owner may retain a sum equal to 150% of the estimated cost of completing any unfinished items, provided that said unfinished items are listed separately and the estimated cost of completing any unfinished items likewise listed separately. Thereafter, Owner shall pay to Construction Manager, monthly, the amount retained for incomplete items as each of said items is completed.

11.3 The Construction Manager shall promptly pay all the amount due Trade Contractors or other persons with whom he has a contract upon receipt of any payment from the Owner, the application for which includes amounts due such Trade Contractor or other persons. Before issuance of final payment, the Construction Manager shall submit satisfactory evidence that all payrolls, materials bills and other indebtedness connected with the Project have been paid or otherwise satisfied.

11.4 If the Owner should fail to pay the Construction Manager within seven (7) days after the time the payment of any amount becomes due, then the Construction Manager may, upon seven (7) additional days' written notice to the Owner and the Architect/Engineer, stop the Project until payment of the amount owing has been received.

11.5 Payments due but unpaid shall bear interest at the rate the Owner is paying on his construction loan or at the legal rate, whichever is higher.

ARTICLE 12

Insurance, Indemnity and Waiver of Subrogation

12.1 Indemnity

12.1.1 The Construction Manager agrees to indemnify and hold the Owner harmless from all claims for bodily injury and property damage (other than the work itself and other property insured under Paragraph 12.4) that may arise from the Construction Manager's operations under this Agreement.

12.1.2 The Owner shall cause any other contractor who may have a contract with the Owner to perform construction or installation work in the areas where work will be performed under this Agreement, to agree to indemnify the Owner and the Construction Manager and hold them harmless from all claims for bodily injury and property damage (other than property insured under Paragraph 12.4) that may arise from that contractor's operations. Such provisions shall be in a form satisfactory to the Construction Manager.

12.2 Construction Manager's Liability Insurance

12.2.1 The Construction Manager shall purchase and maintain such insurance as will protect him from the claims set forth below which may arise out of or result from the Construction Manager's operations under this Agreement whether such operations be by himself or by any Trade Contractor or by anyone directly or indirectly employed by any of them, or by anyone for whose acts any of them may be liable:

12.2.1.1 Claims under workers' compensation, disability benefit and other similar employee benefit acts which are applicable to the work to be performed.

12.2.1.2 Claims for damages because of bodily injury, occupational sickness or disease, or death of his employees under any applicable employer's liability law.

12.2.1.3 Claims for damages because of bodily injury, or death of any person other than his employees.

12.2.1.4 Claims for damages insured by usual personal injury liability coverage which are sustained (1) by any person as a result of an offense directly or indirectly related to the employment of such person by the Construction Manager or (2) by any other person.

12.2.1.5 Claims for damages, other than to the work itself, because of injury to or destruction of tangible property, including loss of use therefrom.

12.2.1.6 Claims for damages because of bodily injury or death of any person or property damage arising out of the ownership, maintenance or use of any motor vehicle.

12.2.2 The Construction Manager's Comprehensive General Liability Insurance shall include premises – operations (including explosion, collapse and underground coverage) elevators, independent contractors, completed operations, and blanket contractual liability on all written contracts, all including broad form property damage coverage.

12.2.3 The Construction Manager's Comprehensive General and Automobile Liability Insurance, as required by Subparagraphs 12.2.1 and 12.2.2 shall be written for not less than limits of liability as follows:

a. Comprehensive General Liability
   1. Personal Injury $_____ Each Occurrence

                          $_____ Aggregate
                                     (Completed Operations)
   2. Property Damage $_____ Each Occurrence

                          $_____ Aggregate

b. Comprehensive Automobile Liability
   1. Bodily Injury $_____ Each Person

                          $_____ Each Occurrence

   2. Property Damage $_____ Each Occurrence

12.2.4 Comprehensive General Liability Insurance may be arranged under a single policy for the full limits required or by a combination of underlying policies with the balance provided by an Excess or Umbrella Liability policy.

12.2.5 The foregoing policies shall contain a provision that coverages afforded under the policies will not be cancelled or not renewed until at least sixty (60) days' prior written notice has been given to the Owner. Certificates of Insurance showing such coverages to be in Force shall be filed with the Owner prior to commencement of the Work.

12.3 Owner's Liability Insurance

12.3.1 The Owner shall be responsible for purchasing and maintaining his own liability insurance and, at his option, may

AGC Standard CM Contract

purchase and maintain such insurance as will protect him against claims which may arise from operations under this Agreement.

12.4 Insurance to Protect Project

12.4.1 The Owner shall purchase and maintain property insurance in a form acceptable to the Construction Manager upon the entire Project for the full cost of replacement as of the time of any loss. This insurance shall include as named insureds the Owner, the Construction Manager, Trade Contractors and their Trade Subcontractors and shall insure against loss from the perils of Fire, Extended Coverage, and shall include "All Risk" insurance for physical loss or damage including without duplication of coverage at least theft, vandalism, malicious mischief, transit, collapse, flood, earthquake, testing, and damage resulting from defective design, workmanship or material. The Owner will increase limits of coverage, if necessary, to reflect estimated replacement cost. The Owner will be responsible for any co-insurance penalties or deductibles. If the Project covers an addition to or is adjacent to an existing building, the Construction Manager, Trade Contractors and their Trade Subcontractors shall be named as additional insureds under the Owner's Property Insurance covering such building and its contents.

12.4.1.1 If the Owner finds it necessary to occupy or use a portion or portions of the Project prior to Substantial Completion thereof, such occupancy shall not commence prior to a time mutually agreed to by the Owner and Construction Manager and to which the insurance company or companies providing the property insurance have consented by endorsement to the policy or policies. This insurance shall not be cancelled or lapsed on account of such partial occupancy. Consent of the Construction Manager and of the insurance company or companies to such occupancy or use shall not be unreasonably withheld.

12.4.2 The Owner shall purchase and maintain such boiler and machinery insurance as may be required or necessary. This insurance shall include the interests of the Owner, the Construction Manager, Trade Contractors and their Subcontractors in the Work.

12.4.3 The Owner shall purchase and maintain such insurance as will protect the Owner and Construction Manager against loss of use of Owner's property due to those perils insured pursuant to Subparagraph 12.4.1. Such policy will provide coverage for expediting expenses of materials, continuing overhead of the Owner and Construction Manager, necessary labor expense including overtime, loss of income by the Owner and other determined exposures. Exposures of the Owner and the Construction Manager shall be determined by mutual agreement and separate limits of coverage fixed for each item.

12.4.4 The Owner shall file a copy of all policies with the Construction Manager before an exposure to loss may occur. Copies of any subsequent endorsements will be furnished to the Construction Manager. The Construction Manager will be given sixty (60) days notice of cancellation, non-renewal, or any endorsements restricting or reducing coverage. If the Owner does not intend to purchase such insurance, he shall inform the Construction Manager in writing prior to the commencement of the Work. The Construction Manager may then effect insurance which will protect the interest of himself, the Trade Contractors and their Trade Subcontractors in the Project, the cost of which shall be a Cost of the Project pursuant to Article 8, and the Guaranteed Maximum Price shall be increased by Change Order. If the Construction Manager is damaged by failure of the Owner to purchase or maintain such insurance or to so notify the Construction Manager, the Owner shall bear all reasonable costs properly attributable thereto.

12.5 Property Insurance Loss Adjustment

12.5.1 Any insured loss shall be adjusted with the Owner and the Construction Manager and made payable to the Owner and Construction Manager as trustees for the insureds, as their interests may appear, subject to any applicable mortgagee clause.

12.5.2 Upon the occurrence of an insured loss, monies received will be deposited in a separate account and the trustees shall make distribution in accordance with the agreement of the parties in interest, or in the absence of such agreement, in accordance with an arbitration award pursuant to Article 16. If the trustees are unable to agree on the settlement of the loss, such dispute shall also be submitted to arbitration pursuant to Article 16.

12.6 Waiver of Subrogation

12.6.1 The Owner and Construction Manager waive all rights against each other, the Architect/Engineer, Trade Contractors, and their Trade Subcontractors for damages caused by perils covered by insurance provided under Paragraph 12.4, except such rights as they may have to the proceeds of such insurance held by the Owner and Construction Manager as trustees. The Construction Manager shall require similar waivers from all Trade Contractors and their Trade Subcontractors.

12.6.2 The Owner and Construction Manager waive all rights against each other and the Architect/Engineer, Trade Contractors and their Trade Subcontractors for loss or damage to any equipment used in connection with the Project and covered by any property insurance. The Construction Manager shall require similar waivers from all Trade Contractors and their Trade Subcontractors.

12.6.3 The Owner waives subrogation against the Construction Manager, Architect/Engineer, Trade Contractors, and their Trade Subcontractors on all property and consequential loss policies carried by the Owner on adjacent properties and under property and consequential loss policies purchased for the Project after its completion.

12.6.4 If the policies of insurance referred to in this Paragraph require an endorsement to provide for continued coverage where there is a waiver of subrogation, the owners of such policies will cause them to be so endorsed.

ARTICLE 13

Termination of the Agreement and Owner's Right to Perform Construction Manager's Obligations

13.1 Termination by the Construction Manager

13.1.1 If the Project is stopped for a period of thirty days under an order of any court or other public authority having jurisdiction, or as a result of an act of government, such as a declaration of a national emergency making materials unavailable, through no act or fault of the Construction Manager, or if the Project should be stopped for a period of thirty days by the Construction Manager for the Owner's failure to make payment thereon, then the Construction Manager may, upon seven days' written notice to the Owner and the Architect/Engineer, terminate this Agreement and recover from the Owner payment for all work executed, the Construction Manager's Fee earned to date, and for any proven loss sustained upon any materials, equipment, tools, construction equipment and machinery, including reasonable profit and damages.

13.2 Owner's Right to Perform Construction Manager's Obligations and Termination by the Owner for Cause

13.2.1 If the Construction Manager fails to perform any of his obligations under this Agreement including any obligation he assumes to perform work with his own forces, the Owner may, after seven days' written notice during which period the Construction Manager fails to perform such obligation, make good such deficiencies. The Guaranteed Maximum Price, if any, shall be reduced by the cost to the Owner of making good such deficiencies.

13.2.2 If the Construction Manager is adjudged a bankrupt, or if he makes a general assignment for the benefit of his creditors, or if a receiver is appointed on account of his insolvency, or if he persistently or repeatedly refuses or fails, except in cases for which extension of time is provided, to supply enough properly skilled workmen or proper materials, or if he fails to make prompt payment to Trade Contractors or for materials or labor, or persistently disregards laws, ordinances, rules, regulations or orders of any public authority having jurisdiction, or otherwise is guilty of a substantial violation of a provision of the Agreement, then the Owner may, without prejudice to any right or remedy and after giving the Construction Manager and his surety, if any, seven days' written notice, during which period Construction Manager fails to cure the violation, terminate the employment of the Construction Manager and take possession of the site and of all materials, equipment, tools, construction equipment and machinery thereon owned by the Construction Manager and may finish the Project by whatever method he may deem expedient. In such case, the Construction Manager shall not be entitled to receive any further payment until the Project is finished nor shall he be relieved from his obligations assumed under Article 6.

13.3 Termination by Owner Without Cause

AGC Standard CM Contract

13.3.1 If the Owner terminates this Agreement other than pursuant to Subparagraph 13.2.2 or Subparagraph 13.3.2, he shall reimburse the Construction Manager for any unpaid Cost of the Project due him under Article 8, plus (1) the unpaid balance of the Fee computed upon the Cost of the Project to the date of termination at the rate of the percentage named in Subparagraph 7.2.1 or if the Construction Manager's Fee be stated as a fixed sum, such an amount as will increase the payment on account of his fee to a sum which bears the same ratio to the said fixed sum as the Cost of the Project at the time of termination bears to the adjusted Guaranteed Maximum Price, if any, otherwise to a reasonable estimated Cost of the Project when completed. The Owner shall also pay to the Construction Manager fair compensation, either by purchase or rental at the election of the Owner, for any equipment retained. In case of such termination of the Agreement the Owner shall further assume and become liable for obligations, commitments and unsettled claims that the Construction Manager has previously undertaken or incurred in good faith in connection with said Project. The Construction Manager shall, as a condition of receiving the payments referred to in this Article 13, execute and deliver all such papers and take all such steps, including the legal assignment of his contractual rights, as the Owner may require for the purpose of fully vesting in him the rights and benefits of the Construction Manager under such obligations or commitments.

13.3.2 After the completion of the Design Phase, if the final cost estimates make the Project no longer feasible from the standpoint of the Owner, the Owner may terminate this Agreement and pay the Construction Manager his Fee in accordance with Subparagraph 7.1.1 plus any costs incurred pursuant to Article 9.

ARTICLE 14

Assignment and Governing Law

14.1 Neither the Owner nor the Construction Manager shall assign his interest in this Agreement without the written consent of the other except as to the assignment of proceeds.

14.2 This Agreement shall be governed by the law of the place where the Project is located.

ARTICLE 15

Miscellaneous Provisions

ARTICLE 16

Arbitration

16.1 All claims, disputes and other matters in question arising out of, or relating to, this Agreement or the breach thereof, except with respect to the Architect/Engineer's decision on matters relating to artistic effect, and except for claims which have been waived by the making or acceptance of final payment shall be decided by arbitration in accordance with the Construction Industry Arbitration Rules of the American Arbitration Association then obtaining unless the parties mutually agree otherwise. This Agreement to arbitrate shall be specifically enforceable under the prevailing arbitration law.

16.2 Notice of the demand for arbitration shall be filed in writing with the other party to this Agreement and with the American Arbitration Association. The demand for arbitration shall be made within a reasonable time after the claim, dispute or other matter in question has arisen, and in no event shall it be made after the date when institution of legal or equitable proceedings based on such claim, dispute or other matter in question would be barred by the applicable statute of limitations.

16.3 The award rendered by the arbitrators shall be final and judgment may be entered upon it in accordance with applicable law in any court having jurisdiction thereof.

16.4 Unless otherwise agreed in writing, the Construction Manager shall carry on the Work and maintain the Contract Completion Date during any arbitration proceedings, and the Owner shall continue to make payments in accordance with this Agreem˙nt.

16.5 All claims which are related to or dependent upon each other, shall be heard by the same arbitrator or arbitrators even though the parties are not the same unless a specific contract prohibits such consolidation.

This Agreement executed the day and year first written above.

ATTEST: OWNER:

ATTEST: CONSTRUCTION MANAGER:

APPENDIX C

Interest Tables

1% COMPOUND INTEREST FACTORS

PERIOD	COMPOUND AMOUNT OF 1	PRESENT WORTH OF 1	COMPOUND AMOUNT OF 1 PER PERIOD	UNIFORM SERIES THAT AMOUNTS TO 1	PRESENT WORTH OF 1 PER PERIOD	UNIFORM SERIES THAT 1 WILL BUY
	$(1+i)^n$	$\dfrac{1}{(1+i)^n}$	$\dfrac{(1+i)^n - 1}{i}$	$\dfrac{i}{(1+i)^n - 1}$	$\dfrac{(1+i)^n - 1}{i(1+i)^n}$	$\dfrac{i(1+i)^n}{(1+i)^n - 1}$
1	1·0100	·99009	1·000	1·00000	·9900	1·01000
2	1·0201	·98029	2·010	·49751	1·9703	·50751
3	1·0303	·97059	3·030	·33002	2·9409	·34002
4	1·0406	·96098	4·060	·24628	3·9019	·25628
5	1·0510	·95146	5·101	·19603	4·8534	·20603
6	1·0615	·94204	6·152	·16254	5·7954	·17254
7	1·0721	·93271	7·213	·13862	6·7281	·14862
8	1·0828	·92348	8·285	·12069	7·6516	·13069
9	1·0936	·91433	9·368	·10674	8·5660	·11674
10	1·1046	·90528	10·462	·09558	9·4713	·10558
11	1·1156	·89623	11·566	·08645	10·3676	·09645
12	1·1268	·88744	12·682	·07884	11·2550	·08884
13	1·1380	·87866	13·809	·07241	12·1337	·08241
14	1·1494	·86996	14·947	·06690	13·0037	·07690
15	1·1609	·86134	16·096	·06212	13·8650	·07212
16	1·1725	·85282	17·257	·05794	14·7178	·06794
17	1·1843	·84437	18·430	·05425	15·5622	·06425
18	1·1961	·83601	19·614	·05098	16·3982	·06098
19	1·2081	·82773	20·810	·04805	17·2260	·05805
20	1·2201	·81954	22·019	·04541	18·0455	·05541
21	1·2323	·81143	23·239	·04303	18·8569	·05303
22	1·2447	·80339	24·471	·04086	19·6603	·05086
23	1·2571	·79544	25·716	·03888	20·4558	·04888
24	1·2697	·78756	26·973	·03707	21·2433	·04707
25	1·2824	·77976	28·243	·03540	22·0231	·04540
26	1·2952	·77204	29·525	·03386	22·7952	·04386
27	1·3082	·76440	30·820	·03244	23·5596	·04244
28	1·3212	·75683	32·129	·03112	24·3164	·04112
29	1·3345	·74934	33·450	·02989	25·0657	·03989
30	1·3478	·74192	34·784	·02874	25·8077	·03874
35	1·4166	·70591	41·660	·02400	29·4085	·03400
40	1·4888	·67165	48·886	·02045	32·8346	·03045
45	1·5648	·63905	56·481	·01770	36·0945	·02770
50	1·6446	·60803	64·463	·01551	39·1961	·02551
55	1·7285	·57852	72·852	·01372	42·1471	·02372
60	1·8166	·55044	81·669	·01224	44·9550	·02224
65	1·9093	·52373	90·936	·01099	47·6266	·02099
70	2·0067	·49831	100·676	·00993	50·1685	·01993
75	2·1091	·47412	110·912	·00901	52·5870	·01901
80	2·2167	·45111	121·671	·00821	54·8882	·01821
85	2·3297	·42922	132·978	·00751	57·0776	·01751
90	2·4486	·40839	144·863	·00690	59·1608	·01690
95	2·5735	·38857	157·353	·00635	61·1429	·01635
100	2·7048	·36971	170·481	·00586	63·0288	·01586

2% COMPOUND INTEREST FACTORS

PERIOD	COMPOUND AMOUNT OF 1	PRESENT WORTH OF 1	COMPOUND AMOUNT OF 1 PER PERIOD	UNIFORM SERIES THAT AMOUNTS TO 1	PRESENT WORTH OF 1 PER PERIOD	UNIFORM SERIES THAT 1 WILL BUY
	$(1+i)^n$	$\dfrac{1}{(1+i)^n}$	$\dfrac{(1+i)^n - 1}{i}$	$\dfrac{i}{(1+i)^n - 1}$	$\dfrac{(1+i)^n - 1}{i(1+i)^n}$	$\dfrac{i(1+i)^n}{(1+i)^n - 1}$
1	1·0200	·98039	1·000	1·00000	·9803	1·02000
2	1·0403	·96116	2·019	·49505	1·9415	·51505
3	1·0612	·94232	3·060	·32675	2·8838	·34675
4	1·0824	·92384	4·121	·24262	3·8077	·26262
5	1·1040	·90573	5·204	·19215	4·7134	·21215
6	1·1261	·88797	6·308	·15852	5·6014	·17852
7	1·1486	·87056	7·434	·13451	6·4719	·15451
8	1·1716	·85349	8·582	·11650	7·3254	·13650
9	1·1950	·83675	9·754	·10251	8·1622	·12251
10	1·2189	·82034	10·949	·09132	8·9825	·11132
11	1·2433	·80426	12·168	·08217	9·7868	·10217
12	1·2682	·78849	13·412	·07455	10·5753	·09455
13	1·2936	·77303	14·680	·06811	11·3483	·08811
14	1·3194	·75787	15·973	·06260	12·1062	·08260
15	1·3458	·74301	17·293	·05782	12·8492	·07782
16	1·3727	·72844	18·639	·05365	13·5777	·07365
17	1·4002	·71416	20·012	·04996	14·2918	·06996
18	1·4282	·70015	21·412	·04670	14·9920	·06670
19	1·4568	·68643	22·840	·04378	15·6784	·06378
20	1·4859	·67297	24·297	·04115	16·3514	·06115
21	1·5156	·65977	25·783	·03878	17·0112	·05878
22	1·5459	·64683	27·298	·03663	17·6580	·05663
23	1·5768	·63415	28·844	·03460	18·2922	·05466
24	1·6084	·62172	30·421	·03287	18·9139	·05287
25	1·6406	·60953	32·030	·03122	19·5234	·05122
26	1·6734	·59757	33·670	·02969	20·1210	·04969
27	1·7068	·58586	35·344	·02829	20·7068	·04829
28	1·7410	·57437	37·051	·02698	21·2812	·04698
29	1·7758	·56311	38·792	·02577	21·8443	·04577
30	1·8113	·55207	40·568	·02464	22·3964	·04464
35	1·9998	·50002	49·994	·02000	24·9986	·04000
40	2·2080	·45289	60·401	·01655	27·3554	·03655
45	2·4378	·41019	71·892	·01390	29·4901	·03390
50	2·6915	·37152	84·579	·01182	31·4236	·03182
55	2·9717	·33650	98·586	·01014	33·1747	·03014
60	3·2810	·30478	114·051	·00876	34·7608	·02876
65	3·6225	·27605	131·126	·00762	36·1974	·02762
70	3·9995	·25002	149·977	·00666	37·4986	·02666
75	4·4158	·22645	170·791	·00585	38·6771	·02585
80	4·8754	·20510	193·771	·00516	39·7445	·02516
85	5·3828	·18577	219·143	·00456	40·7112	·02456
90	5·9431	·16826	247·156	·00404	41·5869	·02404
95	6·5616	·15239	278·084	·00359	42·3800	·02359
100	7·2446	·13803	312·232	·00320	43·0983	·02320

3% COMPOUND INTEREST FACTORS

PERIOD	COMPOUND AMOUNT OF 1	PRESENT WORTH OF 1	COMPOUND AMOUNT OF 1 PER PERIOD	UNIFORM SERIES THAT AMOUNTS TO 1	PRESENT WORTH OF 1 PER PERIOD	UNIFORM SERIES THAT 1 WILL BUY
	$(1+i)^n$	$\dfrac{1}{(1+i)^n}$	$\dfrac{(1+i)^n - 1}{i}$	$\dfrac{i}{(1+i)^n - 1}$	$\dfrac{(1+i)^n - 1}{i(1+i)^n}$	$\dfrac{i(1+i)^n}{(1+i)^n - 1}$
1	1·0300	·97087	1·000	1·00000	·9708	1·03000
2	1·0608	·94259	2·029	·49261	1·9134	·52261
3	1·0927	·91514	3·090	·32353	2·8286	·35353
4	1·1255	·88848	4·183	·23902	3·7170	·26902
5	1·1592	·86260	5·309	·18835	4·5797	·21835
6	1·1940	·83748	6·468	·15459	5·4171	·18459
7	1·2298	·81309	7·662	·13050	6·2302	·16050
8	1·2667	·78940	8·892	·11245	7·0196	·14245
9	1·3047	·76641	10·159	·09843	7·7861	·12843
10	1·3439	·74409	11·463	·08723	8·5302	·11723
11	1·3842	·72242	12·807	·07807	9·2526	·10807
12	1·4257	·70137	14·192	·07046	9·9540	·10046
13	1·4685	·68095	15·617	·06402	10·6349	·09402
14	1·5125	·66111	17·086	·05852	11·2960	·08852
15	1·5579	·64186	18·598	·05376	11·9379	·08376
16	1·6047	·62316	20·156	·04961	12·5611	·07961
17	1·6528	·60501	21·761	·04595	13·1661	·07595
18	1·7024	·58739	23·414	·04270	13·7535	·07270
19	1·7535	·57028	25·116	·03981	14·3237	·06981
20	1·8061	·55367	26·870	·03721	14·8774	·06721
21	1·8602	·53754	28·676	·03487	15·4150	·06487
22	1·9161	·52189	30·536	·03274	15·9369	·06274
23	1·9735	·50669	32·452	·03081	16·4436	·06081
24	2·0327	·49193	34·426	·02904	16·9355	·05904
25	2·0937	·47760	36·459	·02742	17·4131	·05742
26	2·1565	·46369	38·553	·02593	17·8768	·05593
27	2·2212	·45018	40·709	·02456	18·3270	·05456
28	2·2879	·43707	42·930	·02329	18·7641	·05329
29	2·3565	·42434	45·218	·02211	19·1884	·05211
30	2·4272	·41198	47·575	·02101	19·6004	·05101
35	2·8138	·35538	60·462	·01653	21·4872	·04653
40	3·2620	·30655	75·401	·01326	23·1147	·04326
45	3·7815	·26443	92·719	·01078	24·5187	·04078
50	4·3839	·22810	112·796	·00886	25·7297	·03886
55	5·0821	·19676	136·071	·00734	26·7744	·03734
60	5·8916	·16973	163·053	·00613	27·6755	·03613
65	6·8299	·14641	194·332	·00514	28·4528	·03514
70	7·9178	·12629	230·594	·00433	29·1234	·03433
75	9·1789	·10894	272·630	·00366	29·7018	·03366
80	10·6408	·09397	321·362	·00311	30·2007	·03311
85	12·3357	·08106	377·856	·00264	30·6311	·03264
90	14·3004	·06992	443·348	·00225	31·0024	·03225
95	16·5781	·06032	519·271	·00192	31·3226	·03192
100	19·2186	·05203	607·287	·00164	31·5989	·03164

Interest Tables

4% COMPOUND INTEREST FACTORS

PERIOD	COMPOUND AMOUNT OF 1	PRESENT WORTH OF 1	COMPOUND AMOUNT OF 1 PER PERIOD	UNIFORM SERIES THAT AMOUNTS TO 1	PRESENT WORTH OF 1 PER PERIOD	UNIFORM SERIES THAT 1 WILL BUY
	$(1+i)^n$	$\dfrac{1}{(1+i)^n}$	$\dfrac{(1+i)^n - 1}{i}$	$\dfrac{i}{(1+i)^n - 1}$	$\dfrac{(1+i)^n - 1}{i(1+i)^n}$	$\dfrac{i(1+i)^n}{(1+i)^n - 1}$
1	1·0400	·96153	1·000	1·00000	·9615	1·04000
2	1·0815	·92455	2·039	·49019	1·8860	·53019
3	1·1248	·88899	3·121	·32034	2·7750	·36034
4	1·1698	·85480	4·246	·23549	3·6298	·27549
5	1·2166	·82192	5·411	·18462	4·4518	·22462
6	1·2653	·79031	6·632	·15076	5·2421	·19076
7	1·3159	·75991	7·898	·12660	6·0020	·16660
8	1·3685	·73069	9·214	·10852	6·7327	·14852
9	1·4233	·70258	10·582	·09449	7·4353	·13449
10	1·4802	·67556	12·006	·08329	8·1108	·12329
11	1·5394	·64958	13·486	·07414	8·7604	·11414
12	1·6010	·62459	15·025	·06655	9·3850	·10655
13	1·6650	·60057	16·626	·06014	9·9856	·10014
14	1·7316	·57747	18·291	·05466	10·5631	·09466
15	1·8009	·55526	20·023	·04994	11·1183	·08994
16	1·8729	·53390	21·824	·04582	11·6522	·08582
17	1·9479	·51337	23·697	·04219	12·1656	·08219
18	2·0258	·49362	25·645	·03899	12·6592	·07899
19	2·1068	·47464	27·671	·03613	13·1339	·07613
20	2·1911	·45638	29·778	·03358	13·5903	·07358
21	2·2787	·43883	31·969	·03128	14·0291	·07128
22	2·3699	·42195	34·247	·02919	14·4511	·06919
23	2·4647	·40572	36·617	·02730	14·8568	·06730
24	2·5633	·39012	39·082	·02558	15·2469	·06558
25	2·6658	·37511	41·645	·02401	15·6220	·06401
26	2·7724	·36068	44·311	·02256	15·9827	·06256
27	2·8833	·34681	47·084	·02123	16·3295	·06123
28	2·9987	·33347	49·967	·02001	16·6630	·06001
29	3·1186	·32065	52·966	·01887	16·9837	·05887
30	3·2433	·30831	56·084	·01783	17·2920	·05783
35	3·9460	·25341	73·652	·01357	18·6646	·05357
40	4·8010	·20828	95·025	·01052	19·7927	·05052
45	5·8411	·17119	121·029	·00826	20·7200	·04826
50	7·1066	·14071	152·667	·00655	21·4821	·04655
55	8·6463	·11565	191·159	·00523	22·1086	·04532
60	10·5196	·09506	237·990	·00420	22·6234	·04420
65	12·7987	·07813	294·968	·00339	23·0466	·04339
70	15·5716	·06421	364·290	·00274	23·3945	·04274
75	18·9452	·05278	448·631	·00222	23·6804	·04222
80	23·0497	·04338	551·244	·00181	23·9153	·04181
85	28·0436	·03565	676·090	·00147	24·1085	·04147
90	34·1193	·02930	827·983	·00120	24·2672	·04120
95	41·5113	·02408	1012·784	·00098	24·3977	·04098
100	50·5049	·01980	1237·623	·00080	24·5049	·04080

5% COMPOUND INTEREST FACTORS

PERIOD	COMPOUND AMOUNT OF 1	PRESENT WORTH OF 1	COMPOUND AMOUNT OF 1 PER PERIOD	UNIFORM SERIES THAT AMOUNTS TO 1	PRESENT WORTH OF 1 PER PERIOD	UNIFORM SERIES THAT 1 WILL BUY
	$(1+i)^n$	$\dfrac{1}{(1+i)^n}$	$\dfrac{(1+i)^n - 1}{i}$	$\dfrac{i}{(1+i)^n - 1}$	$\dfrac{(1+i)^n - 1}{i(1+i)^n}$	$\dfrac{i(1+i)^n}{(1+i)^n - 1}$
1	1·0500	·95238	1·000	1·00000	·9523	1·05000
2	1·1024	·90702	2·049	·48780	1·8594	·53780
3	1·1576	·86383	3·152	·31720	2·7232	·36720
4	1·2155	·82270	4·310	·23201	3·5459	·28201
5	1·2762	·78352	5·525	·18097	4·3294	·23097
6	1·3400	·74621	6·801	·14701	5·0756	·19701
7	1·4071	·71068	8·142	·12281	5·7863	·17281
8	1·4774	·67683	9·549	·10472	6·4632	·15472
9	1·5513	·64460	11·026	·09069	7·1078	·14069
10	1·6288	·61391	12·577	·07950	7·7217	·12950
11	1·7103	·58467	14·206	·07038	8·3064	·12038
12	1·7958	·55683	15·917	·06282	8·8632	·11282
13	1·8856	·53032	17·712	·05645	9·3935	·10645
14	1·9799	·50506	19·598	·05102	9·8986	·10102
15	2·0789	·48101	21·578	·04634	10·3796	·09634
16	2·1828	·45811	23·657	·04226	10·8377	·09226
17	2·2920	·43629	25·840	·03869	11·2740	·08869
18	2·4066	·41552	28·132	·03554	11·6895	·08554
19	2·5269	·39573	30·539	·03274	12·0853	·08274
20	2·6532	·37688	33·065	·03024	12·4622	·08024
21	2·7859	·35894	35·719	·02799	12·8211	·07799
22	2·9252	·34184	38·505	·02597	13·1630	·07597
23	3·0715	·32557	41·430	·02413	13·4885	·07413
24	3·2250	·31006	44·501	·02247	13·7986	·07247
25	3·3863	·29530	47·727	·02095	14·0939	·07095
26	3·5556	·28124	51·113	·01956	14·3751	·06956
27	3·7334	·26784	54·669	·01829	14·6430	·06829
28	3·9201	·25509	58·402	·01712	14·8981	·06712
29	4·1161	·24294	62·322	·01604	15·1410	·06604
30	4·3219	·23137	66·438	·01505	15·3724	·06505
35	5·5160	·18129	90·320	·01107	16·3741	·06107
40	7·0399	·14204	120·799	·00827	17·1590	·05827
45	8·9850	·11129	159·700	·00626	17·7740	·05626
50	11·4673	·08720	209·347	·00477	18·2559	·05477
55	14·6356	·06832	272·712	·00366	18·6334	·05366
60	18·6791	·05353	353·583	·00282	18·9292	·05282
65	23·8398	·04194	456·797	·00218	19·1610	·05218
70	30·4264	·03286	588·528	·00169	19·3246	·05169
75	38·8326	·02575	756·653	·00132	19·4849	·05131
80	49·5614	·02017	971·228	·00102	19·5964	·05102
85	63·2543	·01580	1245·086	·00080	19·6838	·05080
90	80·7303	·01238	1594·607	·00062	19·7522	·05062
95	103·0346	·00970	2040·693	·00049	19·8058	·05049
100	131·5012	·00760	2610·025	·00038	19·8479	·05038

Interest Tables

6% COMPOUND INTEREST FACTORS

PERIOD	COMPOUND AMOUNT OF 1	PRESENT WORTH OF 1	COMPOUND AMOUNT OF 1 PER PERIOD	UNIFORM SERIES THAT AMOUNTS TO 1	PRESENT WORTH OF 1 PER PERIOD	UNIFORM SERIES THAT 1 WILL BUY
	$(1+i)^n$	$\dfrac{1}{(1+i)^n}$	$\dfrac{(1+i)^n - 1}{i}$	$\dfrac{i}{(1+i)^n - 1}$	$\dfrac{(1+i)^n - 1}{i(1+i)^n}$	$\dfrac{i(1+i)^n}{(1+i)^n - 1}$
1	1·0600	·94339	1·000	1·00000	·9433	1·06000
2	1·1235	·88999	2·059	·48543	1·8333	·54543
3	1·1910	·83961	3·183	·31410	2·6730	·37410
4	1·2624	·79209	4·374	·22859	3·4651	·28859
5	1·3382	·74725	5·637	·17739	4·2123	·23739
6	1·4185	·70496	6·975	·14336	4·9173	·20336
7	1·5036	·66505	8·393	·11913	5·5823	·17913
8	1·5938	·62741	9·897	·10103	6·2097	·16103
9	1·6894	·59189	11·491	·08702	6·8016	·14702
10	1·7908	·55839	13·180	·07586	7·3600	·13586
11	1·8982	·52678	14·971	·06679	7·8868	·12679
12	2·0121	·49696	16·869	·05927	8·3838	·11927
13	2·1329	·46883	18·882	·05296	8·8526	·11296
14	2·2609	·44230	21·015	·04748	9·2949	·10758
15	2·3965	·41726	23·275	·04296	9·7122	·10296
16	2·5403	·39364	25·672	·03895	10·1058	·09895
17	2·6927	·37136	28·212	·03544	10·4772	·09544
18	2·8543	·35034	30·905	·03235	10·8276	·09235
19	3·0255	·33051	33·759	·02962	11·1581	·08962
20	3·2071	·31180	36·785	·02718	11·4699	·08718
21	3·3995	·29415	39·992	·02500	11·7640	·08500
22	3·6035	·27750	43·392	·02304	12·0415	·08304
23	3·8197	·26179	46·995	·02127	12·3033	·08127
24	4·0489	·24697	50·815	·01967	12·5503	·07967
25	4·2918	·23299	54·864	·01822	12·7833	·07822
26	4·5493	·21981	59·156	·01690	13·0031	·07690
27	4·8223	·20736	63·705	·01569	13·2105	·07569
28	5·1116	·19563	68·528	·01459	13·4061	·07459
29	5·4183	·18455	73·639	·01357	13·5907	·07357
30	5·7434	·17411	79·058	·01264	13·7648	·07264
35	7·6860	·13010	111·434	·00897	14·4982	·06897
40	10·2857	·09722	154·761	·00646	15·0462	·06646
45	13·7646	·07265	212·743	·00470	15·4558	·06470
50	18·4201	·05428	290·335	·00344	15·7618	·06344
55	24·6503	·04056	394·172	·00253	15·9905	·06253
60	32·9876	·03031	533·128	·00187	16·1614	·06187
65	44·1449	·02265	719·082	·00139	16·2891	·06139
70	59·0759	·01692	967·932	·00103	16·3845	·06103
75	79·0569	·01264	1300·948	·00076	16·4558	·06076
80	105·7959	·00945	1746·599	·00057	16·5091	·06057
85	141·5788	·00706	2342·981	·00042	16·5489	·06042
90	189·4645	·00527	3141·075	·00031	16·5786	·06031
95	253·5462	·00394	4209·103	·00023	16·6009	·06023
100	339·3020	·00294	5638·367	·00017	16·6175	·06017

7% COMPOUND INTEREST FACTORS

PERIOD	COMPOUND AMOUNT OF 1	PRESENT WORTH OF 1	COMPOUND AMOUNT OF 1 PER PERIOD	UNIFORM SERIES THAT AMOUNTS TO 1	PRESENT WORTH OF 1 PER PERIOD	UNIFORM SERIES THAT 1 WILL BUY
	$(1+i)^n$	$\dfrac{1}{(1+i)^n}$	$\dfrac{(1+i)^n-1}{i}$	$\dfrac{i}{(1+i)^n-1}$	$\dfrac{(1+i)^n-1}{i(1+i)^n}$	$\dfrac{i(1+i)^n}{(1+i)^n-1}$
1	1·0700	·93457	1·000	1·00000	·9345	1·07000
2	1·1448	·87343	2·069	·48309	1·8080	·55309
3	1·2250	·81629	3·214	·31105	2·6243	·38105
4	1·3107	·76289	4·439	·22522	3·3872	·29522
5	1·4025	·71298	5·750	·17389	4·1001	·24389
6	1·5007	·66634	7·153	·13979	4·7665	·20979
7	1·6057	·62274	8·654	·11555	5·3892	·18555
8	1·7181	·58200	10·259	·09746	5·9712	·16746
9	1·8384	·54393	11·977	·08348	6·5152	·15348
10	1·9671	·50834	13·816	·07237	7·0235	·14237
11	2·1048	·47509	15·783	·06335	7·4986	·13335
12	2·2521	·44401	17·888	·05590	7·9426	·12590
13	2·4098	·41496	20·140	·04965	8·3576	·11965
14	2·5785	·38781	22·550	·04434	8·7454	·11434
15	2·7590	·36244	25·129	·03979	9·1079	·10979
16	2·9521	·33873	27·888	·03585	9·4466	·10585
17	3·1588	·31657	30·840	·03242	9·7632	·10242
18	3·3799	·29586	33·999	·02941	10·0590	·09941
19	3·6165	·27650	37·378	·02675	10·3355	·09675
20	3·8696	·25841	40·995	·02439	10·5940	·09439
21	4·1405	·24151	44·865	·02228	10·8355	·09228
22	4·4304	·22571	49·005	·02040	11·0612	·09040
23	4·7405	·21094	53·436	·01871	11·2721	·08871
24	5·0723	·19714	58·176	·01718	11·4693	·08718
25	5·4274	·18424	63·249	·01581	11·6535	·08581
26	5·8073	·17219	68·676	·01456	11·8257	·08456
27	6·2138	·16093	74·483	·01341	11·9867	·08342
28	6·6488	·15040	80·697	·01239	12·1371	·08239
29	7·1142	·14056	87·346	·01144	12·2776	·08144
30	7·6122	·13136	94·460	·01058	12·4090	·08058
35	10·6765	·09366	138·236	·00723	12·9476	·07723
40	14·9744	·06678	199·635	·00500	13·3317	·07500
45	21·0024	·04761	285·749	·00349	13·6055	·07349
50	29·4570	·03394	406·528	·00245	13·8007	·07245
55	41·3149	·02420	575·928	·00173	13·9399	·07173
60	57·9464	·07125	813·520	·00122	14·0391	·07122
65	81·2728	·01230	1146·755	·00087	14·1099	·07087
70	113·9893	·00877	1614·134	·00061	14·1603	·07061
75	159·8760	·00625	2269·657	·00044	14·1963	·07044
80	224·2343	·00445	3189·062	·00031	14·2220	·07031
85	314·5002	·00317	4478·575	·00022	14·2402	·07022
90	441·1029	·00226	6287·185	·00015	14·2533	·07015
95	618·6696	·00161	8823·852	·00011	14·2626	·07011
100	867·7162	·00115	12381·661	·00008	14·2692	·07008

Interest Tables

8% COMPOUND INTEREST FACTORS

PERIOD	COMPOUND AMOUNT OF 1	PRESENT WORTH OF 1	COMPOUND AMOUNT OF 1 PER PERIOD	UNIFORM SERIES THAT AMOUNTS TO 1	PRESENT WORTH OF 1 PER PERIOD	UNIFORM SERIES THAT 1 WILL BUY
	$(1+i)^n$	$\dfrac{1}{(1+i)^n}$	$\dfrac{(1+i)^n - 1}{i}$	$\dfrac{i}{(1+i)^n - 1}$	$\dfrac{(1+i)^n - 1}{i(1+i)^n}$	$\dfrac{i(1+i)^n}{(1+i)^n - 1}$
1	1·0800	·92592	1·000	1·00000	·9259	1·08000
2	1·1663	·85733	2·079	·48076	1·7832	·56076
3	1·2597	·79383	3·246	·30803	2·5770	·38803
4	1·3604	·73502	4·506	·22192	3·3121	·30192
5	1·4693	·68058	5·866	·17045	3·9927	·25045
6	1·5868	·63016	7·335	·13631	4·6228	·21631
7	1·7138	·58349	8·922	·11207	5·2063	·19207
8	1·8509	·54026	10·636	·09401	5·7466	·17401
9	1·9990	·50024	12·487	·08007	6·2468	·16007
10	2·1589	·46319	14·486	·06902	6·7100	·14902
11	2·3316	·42888	16·645	·06007	7·1389	·14007
12	2·5181	·39711	18·977	·05269	7·5360	·13269
13	2·7196	·36769	21·495	·04652	7·9037	·12652
14	2·9371	·34046	24·214	·04129	8·2442	·12129
15	3·1721	·31524	27·152	·03682	8·5594	·11682
16	3·4259	·29189	30·324	·03297	8·8513	·11297
17	3·7000	·27026	33·750	·02962	9·1216	·10962
18	3·9960	·25024	37·450	·02670	9·3718	·10670
19	4·3157	·23171	41·446	·02412	9·6035	·10412
20	4·6609	·21454	45·761	·02185	9·8181	·10185
21	5·0338	·19865	50·422	·01983	10·0168	·09983
22	5·4365	·18394	55·456	·01803	10·2007	·09303
23	5·8714	·17031	60·893	·01642	10·3710	·09642
24	6·3411	·15769	66·764	·01497	10·5287	·09497
25	6·8484	·14601	73·105	·01367	10·6747	·09367
26	7·3963	·13520	79·954	·01250	10·8099	·09250
27	7·9880	·12518	87·350	·01144	10·9351	·09144
28	8·6271	·11591	95·338	·01048	11·0510	·09048
29	9·3172	·10732	103·965	·00961	11·1584	·08961
30	10·0626	·09937	113·283	·00882	11·2577	·08882
35	14·7853	·06763	172·316	·00580	11·6545	·08580
40	21·7245	·04603	259·056	·00386	11·9246	·08386
45	31·9204	·03132	386·505	·00258	12·1084	·08258
50	46·9016	·02132	573·770	·00174	12·2334	·08174
55	68·9138	·01451	848·923	·00117	12·3186	·08117
60	101·2570	·00987	1253·213	·00079	12·3765	·08079
65	148·7798	·00672	1847·247	·00054	12·4159	·08054
70	218·6063	·00457	2720·079	·00036	12·4428	·08036
75	321·2045	·00311	4002·556	·00024	12·4610	·08024
80	471·9547	·00211	5886·934	·00016	12·4735	·08016
85	693·4564	·00144	8655·705	·00011	12·4819·	·08011
90	1018·9149	·00098	12723·936	·00007	12·4877	·08007
95	1497·1203	·00066	18701·503	·00005	12·4916	·08005
100	2199·7612	·00045	27484·515	·00003	12·4943	·08003

9% COMPOUND INTEREST FACTORS

PERIOD	COMPOUND AMOUNT OF 1	PRESENT WORTH OF 1	COMPOUND AMOUNT OF 1 PER PERIOD	UNIFORM SERIES THAT AMOUNTS TO 1	PRESENT WORTH OF 1 PER PERIOD	UNIFORM SERIES THAT 1 WILL BUY
	$(1+i)^n$	$\dfrac{1}{(1+i)^n}$	$\dfrac{(1+i)^n-1}{i}$	$\dfrac{i}{(1+i)^n-1}$	$\dfrac{(1+i)^n-1}{i(1+i)^n}$	$\dfrac{i(1+i)^n}{(1+i)^n-1}$
1	1·0900	·91743	1·000	1·00000	·9174	1·09000
2	1·1880	·84168	2·089	·47846	1·7591	·56846
3	1·2950	·77218	3·278	·30505	2·5312	·39505
4	1·4115	·70842	4·573	·21866	3·3297	·30866
5	1·5386	·64993	5·984	·16709	3·8896	·25709
6	1·6771	·59626	7·523	·13291	4·4859	·22291
7	1·8280	·54703	9·200	·10869	5·0329	·19869
8	1·9925	·50186	11·028	·09067	5·5348	·18067
9	2·1718	·46042	13·021	·07679	5·9952	·16679
10	2·3673	·42241	15·192	·06582	6·4176	·15582
11	2·5804	·38753	17·560	·05694	6·8051	·14694
12	2·8126	·35553	20·140	·04965	7·1607	·13965
13	3·0658	·32617	22·953	·04356	7·4869	·13356
14	3·3417	·29924	26·019	·03843	7·7861	·12843
15	3·6424	·27453	29·360	·03405	8·0606	·12405
16	3·9703	·25186	33·003	·03029	8·3125	·12029
17	4·3276	·23107	36·973	·02704	8·5436	·11704
18	4·7171	·21199	41·301	·02421	8·7556	·11421
19	5·1416	·19448	46·018	·02173	8·9501	·11173
20	5·6044	·17843	51·160	·01954	9·1285	·10954
21	6·1088	·16369	56·764	·01761	9·2922	·10761
22	6·6586	·15018	62·873	·01590	9·4424	·10590
23	7·2578	·13778	69·531	·01438	9·5802	·10438
24	7·9110	·12640	76·789	·01302	9·7066	·10302
25	8·6230	·11596	84·700	·01180	9·8225	·10180
26	9·3991	·10639	93·323	·01071	9·9289	·10071
27	10·2450	·09760	102·723	·00973	10·0265	·09973
28	11·1671	·08954	112·968	·00885	10·1161	·09885
29	12·1721	·08215	124·135	·00805	10·1982	·09805
30	13·2676	·07537	136·307	·00733	10·2736	·09733
35	20·4139	·04898	215·710	·00463	10·5668	·09463
40	31·4094	·03183	337·882	·00295	10·7573	·09295
45	48·3272	·02069	525·858	·00190	10·8811	·09190
50	74·3575	·01344	815·083	·00122	10·9616	·09122
55	114·4082	·00874	1260·091	·00079	11·0139	·09079
60	176·0312	·00568	1944·791	·00051	11·0479	·09051
65	270·8459	·00369	2998·288	·00033	11·0700	·09033
70	416·7300	·00239	4619·223	·00021	11·0844	·09021
75	641·1908	·00155	7113·232	·00014	11·0937	·09014
80	986·5515	·00101	10950·572	·00009	11·0998	·09009
85	1517·9319	·00065	16854·798	·00005	11·1037	·09005
90	2335·5264	·00042	25939·182	·00003	11·1063	·09003
95	3593·4969	·00027	39916·632	·00002	11·1080	·09002
100	5529·0406	·00018	61422·673	·00001	11·1091	·09001

Interest Tables

10% COMPOUND INTEREST FACTORS

PERIOD	COMPOUND AMOUNT OF 1	PRESENT WORTH OF 1	COMPOUND AMOUNT OF 1 PER PERIOD	UNIFORM SERIES THAT AMOUNTS TO 1	PRESENT WORTH OF 1 PER PERIOD	UNIFORM SERIES THAT 1 WILL BUY
	$(1+i)^n$	$\dfrac{1}{(1+i)^n}$	$\dfrac{(1+i)^n - 1}{i}$	$\dfrac{i}{(1+i)^n - 1}$	$\dfrac{(1+i)^n - 1}{i(1+i)^n}$	$\dfrac{i(1+i)^n}{(1+i)^n - 1}$
1	1·1000	·90909	1·000	1·00000	·9090	1·10000
2	1·2099	·82644	2·099	·47619	1·7355	·57619
3	1·3309	·75131	3·309	·30211	2·4868	·40211
4	1·4640	·68301	4·640	·21547	3·1698	·31547
5	1·6105	·62092	6·105	·16379	3·7907	·26379
6	1·7715	·56447	7·715	·12960	4·3552	·22960
7	1·9487	·51315	9·487	·10540	4·8684	·20540
8	2·1435	·46650	11·435	·08744	5·3349	·18744
9	2·3579	·42409	13·579	·07364	5·7590	·17364
10	2·5937	·38554	15·937	·06274	6·1445	·16274
11	2·8531	·35049	18·531	·05396	6·4950	·15396
12	3·1384	·31863	21·384	·04676	6·8136	·14676
13	3·4522	·28966	24·522	·04077	7·1033	·14077
14	3·7974	·26333	27·974	·03574	7·3666	·13574
15	4·1772	·23939	31·772	·03147	7·6060	·13147
16	4·5949	·21762	35·949	·02781	7·8237	·12781
17	5·0544	·19784	40·544	·02466	8·0215	·12466
18	5·5599	·17985	45·599	·02193	8·2014	·12193
19	6·1159	·16350	51·159	·01954	8·3649	·11954
20	6·7274	·14864	57·274	·01745	8·5135	·11745
21	7·4002	·13513	64·002	·01562	8·6486	·11562
22	8·1402	·12284	71·402	·01400	8·7715	·11400
23	8·9543	·11167	79·543	·01257	8·8832	·11257
24	9·8497	·10152	88·497	·01129	8·9847	·11129
25	10·8347	·09229	98·347	·01016	9·0770	·11016
26	11·9181	·08390	109·181	·00915	9·1609	·10915
27	13·1099	·07627	121·099	·00825	9·2372	·10825
28	14·4209	·06934	134·209	·00745	9·3065	·10745
29	15·8630	·06303	148·630	·00672	9·3696	·10672
30	17·4494	·05730	164·494	·00607	9·4269	·10607
35	28·1024	·03558	271·024	·00368	9·6441	·10368
40	45·2592	·02209	442·592	·00225	9·7790	·10225
45	72·8904	·01371	718·904	·00139	9·8628	·10139
50	117·3908	·00851	1163·908	·00085	9·9148	·10085
55	189·0591	·00528	1880·591	·00053	9·9471	·10053
60	304·4816	·00328	3034·816	·00032	9·9671	·10032
65	490·3706	·00203	4893·706	·00020	9·9796	·10020
70	789·7468	·00126	7887·468	·00012	9·9873	·10012
75	1271·8952	·00078	12708·952	·00007	9·9921	·10007
80	2048·4000	·00048	20474·000	·00004	9·9951	·10004
85	3298·9687	·00030	32979·687	·00003	9·9969	·10003
90	5313·0221	·00018	53120·221	·00001	9·9981	·10001
95	8556·6753	·00011	85556·753	·00001	9·9988	·10001
100	13780·6110	·00007	137796·110	·00000	9·9992	·10000

11% COMPOUND INTEREST FACTORS

PERIOD	COMPOUND AMOUNT OF 1	PRESENT WORTH OF 1	COMPOUND AMOUNT OF 1 PER PERIOD	UNIFORM SERIES THAT AMOUNTS TO 1	PRESENT WORTH OF 1 PER PERIOD	UNIFORM SERIES THAT 1 WILL BUY
	$(1+i)^n$	$\dfrac{1}{(1+i)^n}$	$\dfrac{(1+i)^n - 1}{i}$	$\dfrac{i}{(1+i)^n - 1}$	$\dfrac{(1+i)^n - 1}{i(1+i)^n}$	$\dfrac{i(1+i)^n}{(1+i)^n - 1}$
1	1·1100	·90090	1·000	1·00000	·9009	1·11000
2	1·2320	·81162	2·109	·47393	1·7125	·58393
3	1·3676	·73119	3·342	·29921	2·4437	·40921
4	1·5180	·65873	4·709	·21232	3·1024	·32232
5	1·6850	·59345	6·227	·16057	3·6958	·27057
6	1·8704	·53464	7·912	·12637	4·2305	·23637
7	2·0761	·48165	9·783	·10221	4·7121	·21221
8	2·3045	·43392	11·859	·08432	5·1461	·19432
9	2·5580	·39092	14·163	·07060	5·5370	·18060
10	2·8394	·35218	16·722	·05980	5·8892	·16980
11	3·1517	·31728	19·561	·05112	6·2065	·16112
12	3·4984	·28584	22·713	·04402	6·4923	·15402
13	3·8832	·25751	26·211	·03815	6·7498	·14815
14	4·3104	·23199	30·094	·03322	6·9818	·14322
15	4·7845	·20900	34·405	·02906	7·1908	·13906
16	5·3108	·18829	39·189	·02551	7·3791	·13551
17	5·8950	·16963	44·500	·02247	7·5487	·13247
18	6·5435	·15282	50·395	·01984	7·7016	·12984
19	7·2633	·13767	56·939	·01756	7·8392	·12756
20	8·0623	·12403	64·202	·01557	7·9633	·12557
21	8·9491	·11174	72·265	·01383	8·0750	·12383
22	9·9335	·10066	81·214	·01231	8·1757	·12231
23	11·0262	·09069	91·147	·01097	8·2664	·12097
24	12·2391	·08170	102·174	·00978	8·3481	·11978
25	13·5854	·07360	114·413	·00874	8·4217	·11874
26	15·0798	·06631	127·998	·00781	8·4880	·11781
27	16·7386	·05974	143·078	·00698	8·5478	·11698
28	18·5798	·05382	159·817	·00625	8·6016	·11625
29	20·6236	·04848	178·397	·00560	8·6501	·11560
30	22·8922	·04368	199·020	·00502	8·6937	·11502
35	38·5748	·02592	341·589	·00292	8·8552	·11292
40	65·0008	·01538	581·825	·00171	8·9510	·11171
45	109·5302	·00912	986·638	·00101	9·0079	·11101
50	184·5647	·00541	1668·770	·00059	9·0416	·11059
55	311·0023	·00321	2818·203	·00035	9·0616	·11035
60	524·0570	·00190	4755·064	·00021	9·0735	·11021
65	883·0665	·00113	8018·787	·00012	9·0806	·11012
70	1488·0185	·00067	13518·350	·00007	9·0847	·11007
75	2507·3976	·00039	22785·432	·00004	9·0872	·11004
80	4225·1109	·00023	38401·008	·00002	9·0887	·11002
85	7119·5571	·00014	64714·155	·00001	9·0896	·11001
90	11996·8680	·00008	109053·340	·00000	9·0901	·11000
95	20215·4180	·00004	183767·430	·00000	9·0904	·11000
100	34064·1570	·00002	309665·060	·00000	9·0906	·11000

Interest Tables

12% COMPOUND INTEREST FACTORS

PERIOD	COMPOUND AMOUNT OF 1	PRESENT WORTH OF 1	COMPOUND AMOUNT OF 1 PER PERIOD	UNIFORM SERIES THAT AMOUNTS TO 1	PRESENT WORTH OF 1 PER PERIOD	UNIFORM SERIES THAT 1 WILL BUY
	$(1+i)^n$	$\dfrac{1}{(1+i)^n}$	$\dfrac{(1+i)^n - 1}{i}$	$\dfrac{i}{(1+i)^n - 1}$	$\dfrac{(1+i)^n - 1}{i(1+i)^n}$	$\dfrac{i(1+i)^n}{(1+i)^n - 1}$
1	1·1200	·89285	1·000	1·00000	·8928	1·12000
2	1·2543	·79719	2·119	·47169	1·6900	·59169
3	1·4049	·71178	3·374	·29634	2·4018	·41634
4	1·5735	·63551	4·779	·20923	3·0373	·32923
5	1·7623	·56742	6·352	·15740	3·6047	·27740
6	1·9738	·50663	8·115	·12322	4·1114	·24322
7	2·2106	·45234	10·089	·09911	4·5637	·21911
8	2·4759	·40388	12·299	·08130	4·9676	·20130
9	2·7730	·36061	14·775	·06767	5·3282	·18767
10	3·1058	·32197	17·548	·05698	5·6502	·17698
11	3·4785	·28747	20·654	·04841	5·9376	·16841
12	3·8959	·25667	24·133	·04143	6·1943	·16143
13	4·3634	·22917	28·029	·03567	6·4235	·15567
14	4·8871	·20461	32·392	·03087	6·6281	·15087
15	5·4735	·18269	37·279	·02682	6·8108	·14682
16	6·1303	·16312	42·753	·02339	6·9739	·14339
17	6·8660	·14564	48·883	·02045	7·1196	·14045
18	7·6899	·13003	55·749	·01793	7·2496	·13793
19	8·6127	·11610	63·439	·01576	7·3657	·13576
20	9·6462	·10366	72·052	·01387	7·4694	·13387
21	10·8038	·09255	81·698	·01224	7·5620	·13224
22	12·1003	·08264	92·502	·01081	7·6446	·13081
23	13·5523	·07378	104·602	·00955	7·7184	·12955
24	15·1786	·06588	118·155	·00846	7·7843	·12846
25	17·0000	·05882	133·333	·00749	7·8431	·12749
26	19·0400	·05252	150·333	·00665	7·8956	·12665
27	21·3248	·04689	169·373	·00590	7·9425	·12590
28	23·8838	·04186	190·698	·00524	7·9844	·12524
29	26·7499	·03738	214·582	·00466	8·0218	·12466
30	29·9599	·03337	241·332	·00414	8·0551	·12414
35	52·7996	·01893	431·663	·00231	8·1755	·12231
40	93·0509	·01074	767·091	·00130	8·2437	·12130
45	163·9875	·00609	1358·229	·00073	8·2825	·12073
50	289·0021	·00346	2400·017	·00041	8·3044	·12041
55	509·3204	·00196	4236·003	·00023	8·3169	·12023
60	897·5966	·00111	7471·638	·00013	8·3240	·12013
65	1581·8719	·00063	13173·932	·00007	8·3280	·12007
70	2787·7987	·00035	23223·322	·00004	8·3303	·12004
75	4913·0538	·00020	40933·781	·00002	8·3316	·12002
80	8658·4794	·00011	72145·661	·00001	8·3323	·12001
85	15259·1980	·00006	127151·650	·00000	8·3327	·12000
90	26891·9150	·00003	224090·950	·00000	8·3330	·12000
95	47392·7240	·00002	394931·030	·00000	8·3331	·12000
100	83522·2210	·00001	696010·170	·00000	8·3332	·12000

13% COMPOUND INTEREST FACTORS

PERIOD	COMPOUND AMOUNT OF 1	PRESENT WORTH OF 1	COMPOUND AMOUNT OF 1 PER PERIOD	UNIFORM SERIES THAT AMOUNTS TO 1	PRESENT WORTH OF 1 PER PERIOD	UNIFORM SERIES THAT 1 WILL BUY
	$(1+i)^n$	$\dfrac{1}{(1+i)^n}$	$\dfrac{(1+i)^n-1}{i}$	$\dfrac{i}{(1+i)^n-1}$	$\dfrac{(1+i)^n-1}{i(1+i)^n}$	$\dfrac{i(1+i)^n}{(1+i)^n-1}$
1	1·1300	·88495	1·000	1·00000	·8849	1·13000
2	1·2768	·78314	2·129	·46948	1·6681	·59948
3	1·4428	·69305	3·406	·29352	2·3611	·42352
4	1·6304	·61331	4·849	·20619	2·9744	·33619
5	1·8424	·54275	6·480	·15431	3·5172	·28431
6	2·0819	·48031	8·322	·12015	3·9975	·25015
7	2·3526	·42506	10·404	·09611	4·4226	·22611
8	2·6584	·37615	12·757	·07838	4·7987	·20838
9	3·0040	·33288	15·415	·06486	5·1316	·19486
10	3·3945	·29458	18·419	·05428	5·4262	·18428
11	3·8358	·26069	21·814	·04584	5·6869	·17584
12	4·3345	·23070	25·650	·03898	5·9176	·16898
13	4·8980	·20416	29·984	·03335	6·1218	·16335
14	5·5347	·18067	34·882	·02866	6·3024	·15866
15	6·2542	·15989	40·417	·02474	6·4623	·15474
16	7·0673	·14149	46·671	·02142	6·6038	·15142
17	7·9860	·12521	53·739	·01860	6·7290	·14860
18	9·0242	·11081	61·725	·01620	6·8399	·14620
19	10·1974	·09806	70·749	·01413	6·9379	·14413
20	11·5230	·08678	80·946	·01235	7·0247	·14235
21	13·0210	·07679	92·469	·01081	7·1015	·14081
22	14·7138	·06796	105·490	·00947	7·1695	·13947
23	16·6266	·06014	120·204	·00831	7·2296	·13831
24	18·7880	·05322	136·831	·00730	7·2828	·13730
25	21·2305	·04710	155·619	·00642	7·3299	·13642
26	23·9905	·04168	176·850	·00565	7·3716	·13565
27	27·1092	·03688	200·840	·00497	7·4085	·13497
28	30·6334	·03264	227·949	·00438	7·4411	·13438
29	34·6158	·02888	258·583	·00386	7·4700	·13386
30	39·1158	·02556	293·199	·00341	7·4956	·13341
35	72·0684	·01387	546·680	·00182	7·5855	·13182
40	132·7815	·00753	1013·704	·00098	7·6343	·13098
45	244·6413	·00408	1874·164	·00053	7·6608	·13053
50	450·7358	·00221	3459·506	·00028	7·6752	·13028

14% COMPOUND INTEREST FACTORS

PERIOD	COMPOUND AMOUNT OF 1	PRESENT WORTH OF 1	COMPOUND AMOUNT OF 1 PER PERIOD	UNIFORM SERIES THAT AMOUNTS TO 1	PRESENT WORTH OF 1 PER PERIOD	UNIFORM SERIES THAT 1 WILL BUY
	$(1+i)^n$	$\dfrac{1}{(1+i)^n}$	$\dfrac{(1+i)^n-1}{i}$	$\dfrac{i}{(1+i)^n-1}$	$\dfrac{(1+i)^n-1}{i(1+i)^n}$	$\dfrac{i(1+i)^n}{(1+i)^n-1}$
1	1·1400	·87719	1·000	1·00000	·8771	1·14000
2	1·2995	·76946	2·139	·46728	1·6466	·60728
3	1·4815	·67497	3·439	·29073	2·3216	·43073
4	1·6889	·59208	4·921	·20320	2·9137	·34320
5	1·9254	·51936	6·610	·15128	3·4330	·29128
6	2·1949	·45558	8·535	·11715	3·8886	·25715
7	2·5022	·39963	10·730	·09319	4·2883	·23319
8	2·8525	·35055	13·232	·07557	4·6388	·21557
9	3·2519	·30750	16·085	·06216	4·9463	·20216
10	3·7072	·26974	19·337	·05171	5·2161	·19171
11	4·2262	·23661	23·044	·04339	5·4527	·18339
12	4·8179	·20755	27·270	·03666	5·6602	·17666
13	5·4924	·18206	32·088	·03116	5·8423	·17116
14	6·2613	·15971	37·581	·02660	6·0020	·16660
15	7·1379	·14009	43·842	·02280	6·1421	·16280
16	8·1372	·12289	50·980	·01961	6·2650	·15961
17	9·2764	·10779	59·117	·01691	6·3728	·15691
18	10·5751	·09456	68·394	·01462	6·4674	·15462
19	12·0556	·08294	78·969	·01266	6·5503	·15266
20	13·7434	·07276	91·024	·01098	6·6231	·15098
21	15·6675	·06382	104·768	·00954	6·6869	·14954
22	17·8610	·05598	120·435	·00830	6·7429	·14830
23	20·3615	·04911	138·297	·00723	6·7920	·14723
24	23·2122	·04308	158·658	·00630	6·8351	·14630
25	26·4619	·03779	181·870	·00549	6·8729	·14549
26	30·1665	·03314	208·332	·00480	6·9060	·14480
27	34·3899	·02907	238·499	·00419	6·9351	·14419
28	39·2044	·02550	272·889	·00366	6·9606	·14366
29	44·6931	·02237	312·093	·00320	6·9830	·14320
30	50·9501	·01962	356·786	·00280	7·0026	·14280
35	98·1001	·01019	693·572	·00144	7·0700	·14144
40	188·8834	·00529	1342·024	·00074	7·1050	·14074
45	363·6790	·00274	2590·564	·00038	7·1232	·14038
50	700·2329	·00142	4994·520	·00020	7·1326	·14020

15% COMPOUND INTEREST FACTORS

PERIOD	COMPOUND AMOUNT OF 1	PRESENT WORTH OF 1	COMPOUND AMOUNT OF 1 PER PERIOD	UNIFORM SERIES THAT AMOUNTS TO 1	PRESENT WORTH OF 1 PER PERIOD	UNIFORM SERIES THAT 1 WILL BUY
	$(1+i)^n$	$\dfrac{1}{(1+i)^n}$	$\dfrac{(1+i)^n - 1}{i}$	$\dfrac{i}{(1+i)^n - 1}$	$\dfrac{(1+i)^n - 1}{i(1+i)^n}$	$\dfrac{i(1+i)^n}{(1+i)^n - 1}$
1	1·1500	·86956	1·000	1·00000	·8695	1·15000
2	1·3224	·75014	2·149	·46511	1·6257	·61511
3	1·5208	·65751	3·472	·28797	2·2832	·43797
4	1·7490	·57175	4·993	·20026	2·8549	·35026
5	2·0113	·49717	6·742	·14831	3·3521	·29831
6	2·3130	·43232	8·753	·11423	3·7844	·26423
7	2·6600	·37593	11·066	·09036	4·1604	·24036
8	3·0590	·32690	13·726	·07285	4·4873	·22285
9	3·5178	·28426	16·785	·05957	4·7715	·20957
10	4·0455	·24718	20·303	·04925	5·0187	·19925
11	4·6523	·21394	24·349	·04106	4·2337	·19106
12	5·3502	·18690	29·001	·03448	5·4206	·18448
13	6·1527	·16252	34·351	·02911	5·5831	·17911
14	7·0757	·14132	40·504	·02468	5·7244	·17468
15	8·1370	·12289	47·580	·02101	5·8473	·17101
16	9·3576	·10686	55·717	·01794	5·9542	·16794
17	10·7612	·09292	65·075	·01536	6·0471	·16536
18	12·3754	·08080	75·836	·01318	6·1279	·16318
19	14·2317	·07026	88·211	·01133	6·1982	·16133
20	16·3665	·06110	102·443	·00976	6·2593	·15976
21	18·8215	·05313	118·810	·00841	6·3124	·15841
22	21·6447	·04620	137·631	·00726	6·3586	·15726
23	24·8914	·04017	159·276	·00627	6·3988	·15627
24	28·6251	·03493	184·167	·00542	6·4337	·15542
25	32·9189	·03037	212·793	·00469	6·4641	·15469
26	37·8567	·02641	245·711	·00406	6·4905	·15406
27	43·5353	·02296	283·568	·00352	6·5135	·15352
28	59·0656	·01997	327·104	·00305	6·5335	·15305
29	57·5754	·01736	377·169	·00265	6·5508	·15265
30	66·2117	·01510	434·745	·00230	6·5659	·15230
35	133·1755	·00750	881·170	·00113	6·6166	·15113
40	267·8635	·00373	1779·090	·00056	6·6417	·15056
45	538·7692	·00185	3586·128	·00027	6·6542	·15027
50	1083·6573	·00092	7217·715	·00013	6·6605	·15013

Interest Tables

16% COMPOUND INTEREST FACTORS

PERIOD	COMPOUND AMOUNT OF 1	PRESENT WORTH OF 1	COMPOUND AMOUNT OF 1 PER PERIOD	UNIFORM SERIES THAT AMOUNTS TO 1	PRESENT WORTH OF 1 PER PERIOD	UNIFORM SERIES THAT 1 WILL BUY
	$(1+i)^n$	$\dfrac{1}{(1+i)^n}$	$\dfrac{(1+i)^n - 1}{i}$	$\dfrac{i}{(1+i)^n - 1}$	$\dfrac{(1+i)^n - 1}{i(1+i)^n}$	$\dfrac{i(1+i)^n}{(1+i)^n - 1}$
1	1·1600	·86206	1·000	1·00000	·8620	1·16000
2	1·3455	·74316	2·159	·46296	1·6052	·62296
3	1·5608	·64065	3·505	·28525	2·2458	·44525
4	1·8106	·55229	5·066	·19737	2·7981	·35737
5	2·1003	·47611	6·877	·14540	3·2742	·30540
6	2·4363	·41044	8·977	·11138	3·6847	·27138
7	2·8262	·35382	11·413	·08761	4·0385	·24761
8	3·2784	·30502	14·240	·07022	4·3435	·23022
9	3·8029	·26295	17·518	·05708	4·6065	·21708
10	4·4114	·22668	21·321	·04690	4·8332	·20690
11	5·1172	·19541	25·732	·03886	5·0286	·19886
12	5·9360	·16846	30·850	·03241	5·1971	·19241
13	6·8857	·14522	36·786	·02718	5·3423	·18718
14	7·9875	·12519	43·671	·02289	5·4675	·18289
15	9·2655	·10792	51·659	·01935	5·5754	·17935
16	10·7480	·09304	60·925	·01641	5·6684	·17641
17	12·4676	·08020	71·673	·01395	5·7487	·17395
18	14·4625	·06914	84·140	·01188	5·8178	·17188
19	16·7765	·05960	98·603	·01014	5·8774	·17014
20	19·4607	·05138	115·379	·00866	5·9288	·16866
21	22·5744	·04429	134·840	·00741	5·9731	·16741
22	26·1863	·03818	157·414	·00635	6·0113	·16635
23	30·3762	·03292	183·601	·00544	6·0442	·16544
24	35·2364	·02837	213·977	·00467	6·0726	·16467
25	40·8742	·02446	249·213	·00401	6·0970	·16401
26	47·4141	·02109	290·088	·00344	6·1181	·16344
27	55·0003	·01818	337·502	·00296	6·1363	·16296
28	63·8004	·01567	392·502	·00254	6·1520	·16254
29	74·0085	·01351	456·303	·00219	6·1655	·16219
30	85·8498	·01164	530·311	·00188	6·1771	·16188
35	180·3140	·00554	1120·712	·00089	6·2153	·16089
40	378·7210	·00264	2360·756	·00042	6·2334	·16042
45	795·4436	·00125	4965·272	·00020	6·2421	·16020
50	1670·7033	·00059	10435·645	·00009	6·2462	·16009

17% COMPOUND INTEREST FACTORS

PERIOD	COMPOUND AMOUNT OF 1	PRESENT WORTH OF 1	COMPOUND AMOUNT OF 1 PER PERIOD	UNIFORM SERIES THAT AMOUNTS TO 1	PRESENT WORTH OF 1 PER PERIOD	UNIFORM SERIES THAT 1 WILL BUY
	$(1+i)^n$	$\dfrac{1}{(1+i)^n}$	$\dfrac{(1+i)^n - 1}{i}$	$\dfrac{i}{(1+i)^n - 1}$	$\dfrac{(1+i)^n - 1}{i(1+i)^n}$	$\dfrac{i(1+i)^n}{(1+i)^n - 1}$
1	1·1700	·85470	1·000	1·00000	·8547	1·17000
2	1·3688	·73051	2·169	·46082	1·5852	·63082
3	1·6016	·62437	3·538	·28257	2·2095	·45257
4	1·8738	·53365	5·140	·19453	2·7432	·36453
5	2·1924	·45611	7·014	·14256	3·1993	·31256
6	2·5651	·38983	9·206	·10861	3·5891	·27861
7	3·0012	·33319	11·772	·08494	3·9223	·25494
8	3·5114	·28478	14·773	·06768	4·2071	·23768
9	4·1083	·24340	18·284	·05469	4·4505	·22469
10	4·8068	·20803	22·393	·04465	4·6586	·21465
11	5·6239	·17780	27·199	·03676	4·8364	·20676
12	6·5800	·15197	32·823	·03046	4·9883	·20046
13	7·6986	·12989	39·403	·02537	5·1182	·19537
14	9·0074	·11101	47·102	·02123	5·2292	·19123
15	10·5387	·09488	56·110	·01782	5·3241	·18782
16	12·3303	·08110	66·648	·01500	5·4052	·18500
17	14·4264	·06931	78·979	·01266	5·4746	·18266
18	16·8789	·05924	93·405	·01070	5·5338	·18070
19	19·7483	·05063	110·284	·00906	5·5844	·17906
20	23·1055	·04327	130·032	·00769	5·6277	·17769
21	27·0335	·03699	153·138	·00653	5·6647	·17653
22	31·6292	·03161	180·172	·00555	5·6963	·17555
23	37·0062	·02702	211·801	·00472	5·7233	·17472
24	43·2972	·02309	248·807	·00401	5·7464	·17401
25	50·6578	·01974	292·104	·00342	5·7662	·17342
26	59·2696	·01687	342·762	·00291	5·7831	·17291
27	69·3454	·01442	402·032	·00248	5·7975	·17248
28	81·1342	·01232	471·377	·00212	5·8098	·17212
29	94·9270	·01053	552·511	·00180	5·8203	·17180
30	111·0646	·00900	647·438	·00154	5·8293	·17154
35	243·5033	·00410	1426·490	·00070	5·8581	·17070
40	533·8685	·00187	3134·520	·00031	5·8713	·17031
45	1170·4789	·00085	6879·287	·00014	5·8773	·17014
50	2566·2141	·00038	15089·494	·00006	5·8800	·17006

18% COMPOUND INTEREST FACTORS

PERIOD	COMPOUND AMOUNT OF 1	PRESENT WORTH OF 1	COMPOUND AMOUNT OF 1 PER PERIOD	UNIFORM SERIES THAT AMOUNTS TO 1	PRESENT WORTH OF 1 PER PERIOD	UNIFORM SERIES THAT 1 WILL BUY
	$(1+i)^n$	$\dfrac{1}{(1+i)^n}$	$\dfrac{(1+i)^n-1}{i}$	$\dfrac{i}{(1+i)^n-1}$	$\dfrac{(1+i)^n-1}{i(1+i)^n}$	$\dfrac{i(1+i)^n}{(1+i)^n-1}$
1	1·1800	·84745	1·000	1·00000	·8474	1·18000
2	1·3923	·71818	2·179	·45871	1·5656	·63871
3	1·6430	·60863	3·572	·27992	2·1742	·45992
4	1·9387	·51578	5·215	·19173	2·6900	·37173
5	2·2877	·43710	7·154	·13977	3·1271	·31977
6	2·6995	·37043	9·441	·10591	3·4976	·28591
7	3·1854	·31392	12·141	·08236	3·8115	·26236
8	3·7588	·26603	15·326	·06524	4·0775	·24524
9	4·4354	·22545	19·085	·05239	4·3030	·23239
10	5·2338	·19106	23·521	·04251	4·4940	·22251
11	6·1759	·16191	28·755	·03477	4·6560	·21477
12	7·2875	·13721	34·931	·02862	4·7932	·20862
13	8·5993	·11628	42·218	·02368	4·9095	·20368
14	10·1472	·09854	50·818	·01967	5·0080	·19967
15	11·9737	·08351	60·965	·01640	5·0915	·19640
16	14·1290	·07077	72·938	·01371	5·1623	·19371
17	16·6722	·05997	87·068	·01148	5·2223	·19148
18	19·6732	·05083	103·740	·00963	5·2731	·18963
19	23·2144	·04307	123·413	·00810	5·3162	·18810
20	27·3930	·03650	146·627	·00681	5·3527	·18681
21	32·3237	·03093	174·020	·00574	5·3836	·18574
22	38·1420	·02621	206·344	·00484	5·4099	·18484
23	45·0076	·02221	244·486	·00409	5·4321	·18409
24	53·1089	·01882	289·494	·00345	5·4509	·18345
25	62·6686	·01595	342·603	·00291	5·4669	·18291
26	73·9489	·01352	405·271	·00246	5·4804	·18246
27	87·2597	·01146	479·220	·00208	5·4918	·18208
28	102·9665	·00971	566·480	·00176	5·5016	·18176
29	121·5005	·00823	669·447	·00149	5·5098	·18149
30	143·3706	·00697	790·947	·00126	5·5168	·18126
35	327·9971	·00304	1816·650	·00055	5·5386	·18055
40	750·3780	·00133	4163·211	·00024	5·5481	·18024
45	1716·6831	·00058	9531·572	·00010	5·5523	·18010
50	3927·3551	·00025	21813·083	·00004	5·5541	·18004

19% COMPOUND INTEREST FACTORS

PERIOD	COMPOUND AMOUNT OF 1 $(1+i)^n$	PRESENT WORTH OF 1 $\dfrac{1}{(1+i)^n}$	COMPOUND AMOUNT OF 1 PER PERIOD $\dfrac{(1+i)^n - 1}{i}$	UNIFORM SERIES THAT AMOUNTS TO 1 $\dfrac{i}{(1+i)^n - 1}$	PRESENT WORTH OF 1 PER PERIOD $\dfrac{(1+i)^n - 1}{i(1+i)^n}$	UNIFORM SERIES THAT 1 WILL BUY $\dfrac{i(1+i)^n}{(1+i)^n - 1}$
1	1·1900	·84033	1·000	1·00000	·8403	1·19000
2	1·4160	·70616	2·189	·45662	1·5465	·64662
3	1·6851	·59341	3·606	·27730	2·1399	·46730
4	2·0053	·49866	5·291	·18899	2·6385	·37899
5	2·3863	·41904	7·296	·13705	3·0576	·32705
6	2·8397	·35214	9·682	·10327	3·4097	·29327
7	3·3793	·29591	12·522	·07985	3·7056	·26985
8	4·0213	·24867	15·902	·06288	3·9543	·25288
9	4·7854	·20896	19·923	·05019	4·1633	·24019
10	5·6946	·17560	24·708	·04047	4·3389	·23047
11	6·7766	·14756	30·403	·03289	4·4864	·22289
12	8·0642	·12400	37·180	·02689	4·6105	·21689
13	9·5964	·10420	45·244	·02210	4·7147	·21210
14	11·4197	·08756	54·840	·01823	4·8022	·20823
15	13·5895	·07358	66·260	·01509	4·8758	·20509
16	16·1715	·06183	79·850	·01252	4·9376	·20252
17	19·2441	·05196	96·021	·01041	4·9896	·20041
18	22·9005	·04366	115·265	·00867	5·0333	·19867
19	27·2516	·03669	138·166	·00723	5·0700	·19723
20	32·4294	·03083	165·417	·00604	5·1008	·19604
21	38·5910	·02591	197·847	·00505	5·1267	·19505
22	45·9232	·02177	236·438	·00422	5·1485	·19422
23	54·6487	·01829	282·361	·00354	5·1668	·19354
24	65·0319	·01537	337·010	·00296	5·1822	·19296
25	77·3880	·01292	402·042	·00248	5·1951	·19248
26	92·0917	·01085	479·430	·00208	5·2060	·19208
27	109·5892	·00912	571·522	·00174	5·2151	·19174
28	130·4111	·00766	681·111	·00146	5·2227	·19146
29	155·1893	·00644	811·522	·00123	5·2292	·19123
30	184·6752	·00541	966·711	·00103	5·2346	·19103
35	440·7004	·00226	2314·213	·00043	5·2512	·19043
40	1051·6672	·00095	5529·827	·00018	5·2581	·19018
45	2509·6497	·00039	13203·419	·00007	5·2610	·19007
50	5988·9117	·00016	31515·324	·00003	5·2622	·19003

Interest Tables

20% COMPOUND INTEREST FACTORS

PERIOD	COMPOUND AMOUNT OF 1	PRESENT WORTH OF 1	COMPOUND AMOUNT OF 1 PER PERIOD	UNIFORM SERIES THAT AMOUNTS TO 1	PRESENT WORTH OF 1 PER PERIOD	UNIFORM SERIES THAT 1 WILL BUY
	$(1+i)^n$	$\dfrac{1}{(1+i)^n}$	$\dfrac{(1+i)^n - 1}{i}$	$\dfrac{i}{(1+i)^n - 1}$	$\dfrac{(1+i)^n - 1}{i(1+i)^n}$	$\dfrac{i(1+i)^n}{(1+i)^n - 1}$
1	1·2000	·83333	1·000	1·00000	·8333	1·20000
2	1·4399	·69444	2·199	·45454	1·5277	·65454
3	1·7279	·57870	3·639	·27472	2·1064	·47472
4	2·0735	·48225	5·367	·18628	2·5887	·38628
5	2·4883	·40187	7·441	·13437	2·9906	·33437
6	2·9859	·33489	9·929	·10070	3·3255	·30070
7	3·5831	·27908	12·915	·07742	3·6045	·27742
8	4·2998	·23256	16·499	·06060	3·8371	·26060
9	5·1597	·19380	20·798	·04807	4·0309	·24807
10	6·1917	·16150	25·958	·03852	4·1924	·23852
11	7·4300	·13458	32·150	·03110	4·3270	·23110
12	8·9160	·11215	39·580	·02526	4·4392	·22526
13	10·6993	·09346	48·496	·02062	4·5326	·22062
14	12·8391	·07788	59·195	·01689	4·6105	·21689
15	15·4070	·06490	72·035	·01388	4·6754	·21388
16	18·4884	·05408	87·442	·01143	4·7295	·21143
17	22·1861	·04507	105·930	·00944	4·7746	·20944
18	26·6233	·03756	128·116	·00780	4·8121	·20780
19	31·9479	·03130	154·739	·00646	4·8434	·20646
20	38·3375	·02608	186·687	·00535	4·8695	·20535
21	46·0051	·02173	225·025	·00444	4·8913	·20444
22	55·2061	·01811	271·030	·00368	4·9094	·20368
23	66·2473	·01509	326·236	·00306	4·9245	·20306
24	79·4968	·01257	392·484	·00254	4·9371	·20254
25	95·3961	·01048	471·980	·00211	4·9475	·20211
26	114·4754	·00873	567·377	·00176	4·9563	·20176
27	137·3705	·00727	681·852	·00146	4·9636	·20146
28	164·8446	·00606	819·223	·00122	4·9696	·20122
29	197·8135	·00505	984·067	·00101	4·9747	·20101
30	237·3762	·00421	1181·881	·00084	4·9789	·20084
35	590·6680	·00169	2948·340	·00033	4·9915	·20033
40	1469·7711	·00068	7343·855	·00013	4·9965	·20013
45	3657·2606	·00027	18281·303	·00005	4·9986	·20005
50	9100·4350	·00010	45497·175	·00002	4·9994	·20002

APPENDIX D

Solutions to Exercises

Exercise 1.1

The list of projects that have been completed over their planned time or cost budget is too extensive to list here. Also numerous projects have had quality deficiencies, some of which have resulted in structural or functional failure. Readers should list and discuss projects in their own geographic areas. On a national level, projects that have received attention because of time or cost overruns include the Sidney Opera House in Australia, the New Orleans Superdome, and the King Stadium in Seattle. Projects that have received attention because of their noncompliance with quality standards include the Hartford Civic Center and Yankee Stadium in New York. The time, cost, and quality difficulties of all of these projects have incurred for a combination of reasons. However, the advocate of the CM process would argue that the implementation of the process could have lessened the shortcomings.

Exercise 1.2

Owner: If the CM firm contracts directly with the contractors it can be argued that the CM firm is in the role of a general contractor. Regarding the project owner, this might mean that the CM firm is forced into serving the needs of the contractors as a higher priority than serving its client. Additionally, the fact that the owner can play a more direct role in its choice of contractors and its control of contractors is lost if the contractors are engaged by contract with the CM firm. Most importantly, the CM firm may be better able to serve both its client and the contractor team if it is viewed as independent of both.

Solutions to Exercises

Contractors: One of the advantages of the CM process is that the contractors, through their direct contract with a project owner, cannot be forced into the awkward position of not having direct contract or communication links with the project owner. This benefit is lost if the contractors are engaged directly by the CM firm. Additionally, the contractors may be subject to longer payment delays under this process.

CM Firm: When the CM firm engages contractors directly, it takes on legal liability for their performance. This is true independent of what their contract indicates. This liability can require justification of a larger CM fee or at a minimum indicate a need for significant and costly professional liability coverage for the CM firm. The added liability is especially troublesome if the CM firm cannot obtain authority over the contractors for their performance independent of whether or not it has contract with them.

Exercise 2.1

The determination of the benefits of the CM process to a project is dependent on the characteristics of the specific project. However the following might be considered as an evaluation made only considering the typical project.

	Residential Project	Commercial and Industrial	Heavy and Highway
Systems approach	2	3	1
Improved estimating	1	3	2
Single source of management	2	3	1
Reduces owner's time	1	3	2
Phasing of work	1	2	3
Owner purchased material	2	3	1
Eliminate adverse relations	2	3	1

Note: 3 highest benefit, 1 lowest benefit

EXERCISE 2:2

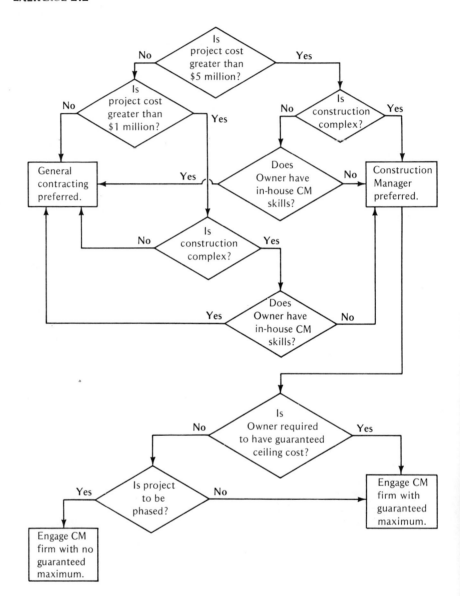

Solutions to Exercises 307

Exercise 3.1

The first article of the AGC contract contains the promise by the construction manager that he accepts the responsibility for the project and that he will use all his abilities in the best interest of the owner. This statement is followed by a brief description of the "construction team" and a clause stipulating the extent of the agreement. (Article eleven of the AIA contract contains the section on extent of the agreement.) The AIA contract contains the same promise, word for word, of acceptance and responsibility. The AIA contract does have a section stating that the construction manager is an agent of the owner, but it fails to bring out the concept of the "construction team."

Article two of the AGC contract deals with the construction manager's services. These services are broken down into two phases. The first being the design phase and the second the construction phase. The AIA contract also breaks down the construction manager's services into these two phases. The first section of the AIA contract also contains a clause that states that compensation will be renegotiated if the scope of the agreement changes materially. Along with this clause is a section for setting forth the time period, stipulated in number of months, it will take to complete the project. Finally, the AIA contract sets out right in the beginning of the contract document the method of compensation that the owner will give the construction manager for his services. The AGC and AIA contract documents break down the services of the construction manager during the design phase and construction phase as:

Design Phase	
AGC	AIA
1. Consultation during project development	1. Consultation during project development
2. Scheduling	2. Scheduling
3. Project budget	3. Project budget
4. Coordination of contract documents	4. Coordination of contract documents
5. Construction planning (includes bidding, labor analysis, and contract awards)	5. Labor (analysis) (includes an equal employment clause)
6. Equal employment opportunity	6. Bidding
	7. Contract awards

Construction Phase

AGC	AIA
1. Project control	1. Project control
2. Cost control	2. Cost control (included are change orders)
3. Change orders	3. Permits and fees
4. Payments to trade contractors	4. Owner's consultants
5. Permits and fees	5. Inspection
6. Owner's consultants	6. Contract performance
7. Inspection (includes a safety review clause)	7. Shop drawings and samples
8. Contract interpretations (same as contract performance)	8. Reports and records
9. Shop drawings and samples	9. Owner-purchased items
10. Reports and project site documents	10. Substantial completion
11. Substantial completion	11. Start-up
12. Start-up	12. Final completion
13. Final completion	13. Additional services

As can be seen from the above table, both contract documents offer the same services in the design phase of the project. The only true difference between the two in this part of the contract is that the AGC contract goes into more detail about the scheduling services than the AIA contract.

In the construction phase both contracts offer basically the same services. The AIA document states specifically when work can be stopped or rejected if not in conformance with the contract documents. Also, the AIA contract offers services in connection with owner-purchased items. The AIA contract does not offer services in the area of project safety, but the AGC contract offers to review contractor's safety programs and make recommendations to the owner. Finally, the AIA contract offers a list of eleven additional services that can be selected by the owner for additional compensation. The AGC contract makes no such offer of additional services.

The third article of the AGC contract deals with the owner's responsibilities. (Owner's responsibility can be found in Article two of the AIA contract.) These two clauses are the same in both contracts, except that in the AIA contract the owner is required to provide insurance counseling services. The AGC contract only requires that legal and accounting counseling services be furnished by the owner, as does the AIA contract.

The fourth article of the AGC contract sets out the procedure for contracting work with trade contractors. The AIA contract has no provisions for this procedure and fails to discuss the matter of trade contractors anywhere in the document.

Solutions to Exercises

Article five of the AGC contract deals with setting up of a contract time schedule. The AIA contract does not have a section for this option.

Article six of the AGC contract allows for a guaranteed maximum price to be set for the project. The AIA contract has no provision for setting a guaranteed maximum price for a project.

Article seven of the AGC contract deals with construction manager's fees. The AGC contract offers a fixed fee for both the design and the construction phase of a project. The AIA contract offers the owner a choice of four methods of compensation. They are:

1. Professional fee plus expenses
2. Multiple of direct personnel expense
3. Fixed fee
4. Percentage of construction costs

The AIA formulas for compensation do not offer a separate fee for the design and the construction phase of a project. What they do offer is for a portion of the total fee to be paid as certain phases of the project are completed. The compensation at the completion of each of the phases must equal 20 percent for the design phase and 100 percent for the construction phase. These percentages are of the total compensation to be paid for basic services. The AIA contract requires that an initial payment be made upon execution of the agreement, which means before any services have been performed. This initial payment is to be negotiated between the owner and construction manager. The AGC contract has no such provision. Finally, in the AIA contract the compensation originally agreed upon is open for renegotiation if the services are not completed in the specified time period. This type of clause is not found in the AGC contract.

Article eight of the AGC contract deals with changes in the work. This sets forth the specific procedures that must be used to obtain compensation for changes in work, types of changes that can be compensated for, and methods of compensation. There is no article in the AIA contract document that deals with changes in the work.

Article nine of the AGC contract deals with cost of work (cost of work is found in Article three of the AIA contract). The AGC contract discusses construction costs as the actual cost of work to the owner. In the AIA contract the term construction cost is used to determine the basis for compensation owed to the construction manager. In the AIA contract the construction manager's responsibilities are undefined and open to interpretation, as far as construction costs are concerned. In the AGC contract a guaranteed maximum price is set forth and all costs are defined in detail. This is done to ensure that the construction manager works to the best of his ability to obtain responsible bids.

Article ten of the AGC contract deals with discounts. These are discounts in relationship to prompt payment of bills. The AIA contract does not discuss the subject of discounts.

Article eleven of the AGC contract deals with payments to the construction manager. (Payments to the construction manager are dealt with in Article six of the AIA contract.) The AGC contract sets up a method of payment for the cost of the work done, plus the construction manager's fee. This method requires monthly reports to be turned in to the owner before any money is paid out. These reports must show detailed accounts of all money paid out, costs accumulated, or costs incurred through the work that had been completed that month. This method also gives provisions for completion of unfinished items before final payment or acceptance of the project. The AIA contract only makes provisions for payment of compensation in terms of services rendered to that date.

Article twelve of the AGC contract deals with insurance. (Insurance is covered in Article ten of the AIA contract.) The AGC contract sets forth a complete insurance program that relates to all of the insurance needs of the construction manager, the owner, and the project. The AIA contract only deals with insurance to be furnished by the construction manager for his protection from claims.

Article thirteen of the AGC contract deals with termination of agreement. (Termination of agreement is dealt with in Article seven of the AIA contract.) The AGC contract gives methods of termination by either the owner or the construction manager. Also determination of cost at termination are specified in the AGC contract. The AIA contract is much less explicit as to how and why a contract can be terminated, or by whom. It basically discusses how the construction manager will be paid in case of termination.

Article fourteen of the AGC contract deals with assignment and governing laws. (These are addressed in Article twelve of the AIA contract.) This subject is covered equally in both contract documents.

Article fifteen of the AGC contract deals with miscellaneous provisions. (This is covered in Article thirteen of the AIA contract.) Both contract documents cover this section equally and make space for this provision.

Article sixteen of the AGC contract deals with arbitration. (Arbitration is covered in Article nine of the AIA contract.) On the subject of arbitration, the two contracts only differ slightly. The AGC contract has an extra clause that states that the construction manager agrees to carry on the work unless otherwise agreed to by him and the owner in writing. The AIA contract has no such clause.

Solutions to Exercises

Exercise 3.2

The ranking of the benefits from CM services is in part dependent on knowledge of the actual project owner and project. However the following represents an evaluation made given the type of owner and project.

	School Board	Group of Doctor Investors	Executives
	Grammar school	Office building	Manufacturing plant
Service			
a. Secure financing	8	3	8
b. Tax analysis of building	10	1	5
c. Value engineering	6	9	1
d. Phasing contract	5	7	3
e. Parameter estimating	4	8	7
f. Owner-purchased material	3	10	6
g. Project planning and scheduling	1	2	2
h. Project supervision	2	4	4
i. Property management	9	6	9
j. Market project	7	5	10

Exercise 4.1

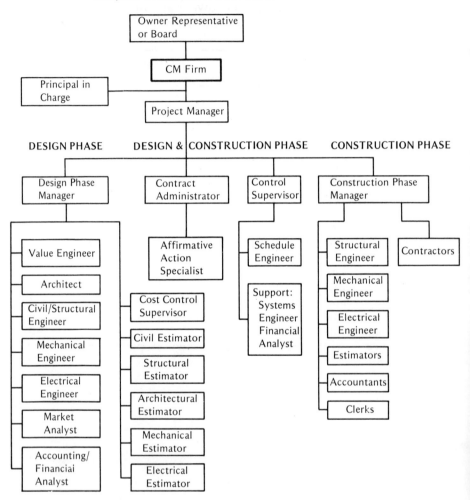

Exercise 4.2

	Weight
Evaluator _____	
Firm _____	
Narrative	5
Project Description	10

Construction Management Control System
  Narrative Reporting
  Master Schedule
  Cost Control Coordination
  Financial/Cash Flow
	25

Major Tasks—Preconstruction	5
Major Tasks—Construction	10
Organization Chart	20

Special Considerations
  Coordination of demo/site preparation
  Job-site Control
  Equal Opportunity
  Inspection
  Employment of Small Business/Minorities
  Pre-bid Qualification
  Guaranteed Schedule
  Impact of Illinois law
	25

Total	100

Exercise 5.1

Pro Forma
First Year Cash
Example Plant Building

Gross Possible Revenue		$250,000
Operating Expenses		−100,000
Net Operating Cash Income		$150,000
Less Mortgage ($1,200,000 × (0.11746)		−140,952
Net Cash Flow		$ 9,048
Depreciation, Interest, and Tax Effects		
Gross possible revenue	$250,000	
Operating expenses	−100,000	
Depreciation		
[$1,200,000 (0.04) (1.5)]	− 72,000	
Interest cost		
($1,200,000) (0.10)	−120,000	
Net Operating Gain (Loss)	$ (42,000)	
Tax Liability ($42,000) (0.40)		16,800
Total Cash Inflow After Taxes		$ 25,848

Exercise 5.2

The life established for each component is dependent on the use of the component as well as its quality of construction. Therefore, the assigned component life will be dependent on the reader's interpretation of use and quality of the component. An example assignment of component lives might be as follows.

Component	Cost	Lives
Storm drains	$ 1,000	50
Building structure roof drains	8,000	50
Service equipment		
Sewage	8,000	50
Plumbing fixtures	10,000	25
Plumbing connections	14,000	25
Water mains	18,000	25
Natural gas mains	6,000	25
Heat exchangers	2,000	15
Air conditioning connector	2,000	15

Solutions to Exercises

Component	Cost	Lives
Processing facilities		
Industrial waste system	60,000	25
Gas process piping	300,000	25
Reflection pool piping	5,000	20
Steam piping	90,000	20
Steam boiler	30,000	20
Process pumps	42,000	15
Laboratory sink	1,000	10
Drinking fountains	1,000	10
Emergency shower wash	1,000	10
Fire hose cabinet	1,000	10
	$600,000	

The calculation and comparison of depreciation lives is as follows:

No Componentizing:
Depreciation rate $= (1.5)(1/40) = 0.0375$
First-year depreciation $= (\$600,000)(0.0375) = \$22,500$

Componentizing:

50-year	25-year	20-year	15-year	10-year
$ 1,000	$ 10,000	$ 5,000	$ 2,000	$1,000
8,000	14,000	90,000	2,000	1,000
8,000	18,000	30,000	42,000	1,000
	6,000			1,000
	60,000			
	300,000			
$17,000	$408,000	$125,000	$46,000	$4,000
×0.03	×0.06	×0.075	×0.10	×0.15
$ 510	$ 24,480	$ 9,375	$ 4,600	$ 600

First year depreciation $= 510 + 24{,}480 + 9{,}375 + 4{,}600 + 600$
First year depreciation $= \$39{,}565$

Exercise 6.1

An example parameter estimating system for estimating the cost of a residential single-family house is as follows:

No.	Parameter
1	Square feet of floor space
2	Lot size
3	Number of rooms
4	Number of floors
5	Square feet of exterior masonry

Trade	Parameter Code
Sitework and landscaping	2
Exterior concrete work	2
Foundation	2
Masonry	5
Exterior carpentry	1
Finish carpentry	1
Roofing and flashing	2
Doors and windows	4
Drywall	3
Painting	1
Finishes	3
Tile work	1
Electrical	1
Plumbing	4
HVAC	4

Exercise 6.2

The factors to consider in attempting to forecast the cost of a specific work package are almost infinite. However, the following are some of the more obvious factors.

Time of the year (i.e. season)
Expected weather
Geographic location
Labor morale
Availability of labor
Supervision attained
Management leadership exhibited
Availability of material

Solutions to Exercises

Harmony with other work phases
Physical environment of surrounding work
Equipment performance
Expected number and severity of disputes
Frequency and severity of accidents

Exercise 7.1

The order of the steps taken by the value engineering team should reflect the order of steps that should yield the largest benefit-cost ratio. The following order of steps is an example.

1. The identification of the owner's functions to be attained from the overall project should be identified and the building layout analyzed for compatibility with the functions.
2. Major building areas, floor space, and buildings wings should be analyzed with the intent of reducing the area, floor space, or wings.
3. Expensive rooms or areas should be analyzed with the intent of cost reductions in the materials used and possible elimination of some fringes.
4. Repetitive items or components in the building should be analyzed. For example, a door design that appears numerous times, floor material used throughout the project, and fixtures such as cabinets should be analyzed. Any cost reduction in a repetitive item or component multiplies several times to sum to a significant savings.
5. Large dollar cost items or building elements should be analyzed with an objective of considering an alternative element that serves the project owner at a lesser cost.

Exercise 7.2

Although the answer to this question is dependent on the actual wall system selected for analysis, the following represents an example solution.

VALUE ENGINEERING

Step 1: Identify Functions of Item Being Analyzed

1. Isolate space
2. Provide aesthetics
3. Insulate noise
4. Insulate temperature
5. Provide durability
6. Hold utility connections
7. Resist lateral loads

VALUE ENGINEERING Continued

Step 2: Make Initial Estimate of Value of Each Function

Function	Initial Estimate of Value
1. Isolate space	40
2. Provide aesthetics	15
3. Insulate noise	10
4. Insulate temperature	10
5. Provide durability	10
6. Hold utility connections	5
7. Resist lateral loads	10

Step 3: List Individual Parts of Item Being Analyzed

A. Paneling
B. Lumber (studs and plates)
C. Nails
D. Running board (bottom)
E. Electrical outlets
F. Running board (top)

Step 4: Identify Labor and Material Cost for Each Part

Part	Labor	Material	Total
A. Paneling	$ 30	$115	$145
B. Lumber (studs & plates)	50	35	85
C. Nails	10	5	15
D. Running board (bottom)	15	15	30
E. Electrical outlets	30	20	50
F. Running board (top)	15	10	25
	$150	$200	$350

Step 5: Identify Function of Each Part

Part	Function	
A. Paneling	1. Isolate space	20%
	2. Provide aesthetics	50%
	4. Insulate temperature	10%
	5. Provide durability	10%
	7. Resist lateral loads	10%
B. Lumber (studs and plates)	1. Isolate space	15%
	3. Insulate noise	15%
	4. Insulate temperature	15%
	6. Hold utility connections	15%
	7. Resist lateral loads	40%
C. Nails	5. Provide durability	25%
	6. Hold utility connections	25%
	7. Resist lateral loads	50%
D. Running Board (bottom)	2. Provide aesthetics	50%
	5. Provide durability	50%
E. Electrical outlets	6. Hold utility connections	100%
F. Running board (top)	2. Provide aesthetics	100%

Solutions to Exercises

VALUE ENGINEERING Continued

Step 6: Calculate Cost per Function

1. Isolate space	A. Paneling	$ 29.00
	B. Lumber (studs and plates)	12.75
	Total	$ 41.75
2. Provide aesthetics	A. Paneling	$ 72.50
	D. Running Board (bottom)	15.00
	F. Running board (top)	25.00
	Total	$112.50
3. Insulate noise	B. Lumber (studs and plates)	$ 12.75
	Total	$ 12.75
4. Insulate temperature	A. Paneling	$ 14.50
	B. Lumber (studs and plates)	12.75
	Total	$ 27.25
5. Provide durability	A. Paneling	$ 14.50
	C. Nails	3.75
	D. Running board (bottom)	15.00
	Total	$ 33.25
6. Hold utility connections	B. Lumber (studs and plates)	$ 12.75
	C. Nails	3.75
	E. Electrical outlets	50.00
	Total	$66.50
7. Resist lateral loads	A. Paneling	$ 14.50
	B. Lumber (studs and plates)	34.00
	C. Nails	7.50
	Total	$ 56.00

Step 7: Create Improved Value

Function	Initial $	Estimate %	Actual $	per VE %
1. Isolate space	140.00	40	41.75	11.9
2. Provide aesthetics	52.50	15	112.50	32.2
3. Insulate noise	35.00	10	12.75	3.6
4. Insulate temperature	35.00	10	27.25	7.7
5. Provide durability	35.00	10	33.25	9.6
6. Hold utility connections	17.50	5	66.50	19.0
7. Resist lateral loads	35.00	10	56.00	16.0
	$350.00	100%	$350.00	100.0%

Appendix D

Exercise 8.1

Work Package	Duration (Months)	EST	EFT	LST	LFT	TF
A	3	0	3	1	4	1
B	2	3	5	17	19	14
C	6	3	9	4	10	1
D	2	0	2	0	2	0
E	7	2	9	2	9	0
F	5	2	7	14	19	12
G	9	9	18	10	19	1
H	10	9	19	9	19	0
I	1	19	20	19	20	0

Exercise 8.2

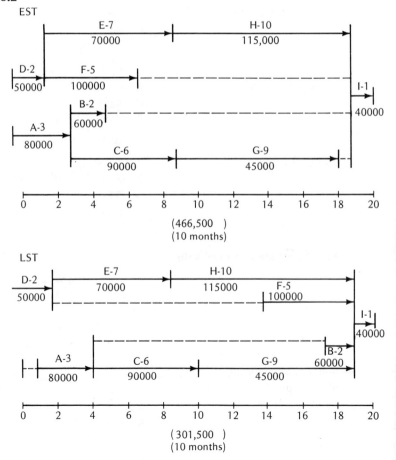

Solutions to Exercises

Exercise 9.1

Included in the potential difficulties or problem areas that can be the result of phasing work are the following:

1. Phasing can result in the need for expensive and time-consuming change orders.
2. Phasing can be restricted or result in problems if the overlapped work requires an excessive amount of resources at any one point in time.
3. When work disputes or delays occur, it may be more difficult for the project owner or its construction manager to pinpoint the cause in a phased versus linear process.
4. Because of the strong dependence of two work processes that are phased, any delay or difficulty with one process is likely to affect immediately the performance of the other, which in turn can cause disputes and discontent among the entities performing the two work processes.
5. Events occur during a construction project that cannot be expected or predicted. Given this fact, the project plan needs constant revision to reflect these events to enable the construction manager to phase work effectively. What to phase and when to phase change as events occur throughout a project.

Exercise 9.2

1. Minimum cost schedule = 37 weeks, $314,000
   - A—1 package, 8 weeks
   - B—3 packages, 11 weeks
   - C—3 packages, 21 weeks
   - D—1 package, 2 weeks
   - E —1 package, 5 weeks
   - F —3 packages, 3 weeks
   - G—2 packages, 10 weeks
   - H—1 package, 6 weeks
   - I —1 package, 2 weeks
   - J —1 package, 5 weeks

2. Minimum time schedule = 32 weeks, $342,000
   - A—3 packages, 5 weeks
   - B—3 packages, 11 weeks
   - C—3 packages, 21 weeks
   - D—1 package, 2 weeks
   - E —1 package, 5 weeks
   - F —3 packages, 3 weeks
   - G—1 package, 8 weeks
   - H—2 packages, 4 weeks
   - I —1 package, 2 weeks
   - J —1 package, 5 weeks

3. Suggested schedule for minimum cost and time = 33 weeks, $326,000
   - A—2 packages, 6 weeks
   - B —3 packages, 11 weeks
   - C—3 packages, 21 weeks
   - D—1 package, 2 weeks
   - E —1 package, 5 weeks
   - F —3 packages, 3 weeks
   - G—2 packages, 10 weeks
   - H—2 packages, 4 weeks
   - I —1 package, 2 weeks
   - J —1 package, 5 weeks

Exercise 10.1

Work Package or Phase	Dollar Contract Amount	Total Construction Billings through June	Complete through June	Earned Revenue through June	Over (Under) Billings	Cash Paid through June	Cash to be Paid in July
Concrete footings	$ 200,000	$200,000	100%	$200,000	$ 0	$200,000	$ 0
Concrete walls	400,000	250,000	50%	200,000	50,000	200,000	200,000
Concrete slabs	500,000	100,000	25%	125,000	(25,000)	100,000	400,000
Concrete beams	100,000	50,000	40%	40,000	10,000	40,000	60,000
	$1,200,000	$600,000		$565,000	$35,000	$540,000	$660,000

$$\% \text{ Complete} = \frac{\text{Earned Revenue}}{\text{Contract Amount}} = \frac{\$565,000}{\$1,200,000} = 47.08\%$$

Solutions to Exercises

Exercise 10.2

<table>
<tr><td colspan="5" align="center">MPDM PROCESSING</td></tr>
<tr><td>Method:</td><td colspan="4">Production Unit:</td></tr>
<tr><td>Units</td><td>Production Total Time</td><td>Number of Cycles</td><td>Mean Cycle Time</td><td>Cycle time — nondelay cycle time</td></tr>
<tr><td>A. Non-delayed Production Cycles</td><td>438</td><td>5</td><td>87.6</td><td>7.6</td></tr>
<tr><td>B. Overall Production Cycles</td><td>1580</td><td>12</td><td>131.66</td><td>47</td></tr>
<tr><td colspan="5" align="center">Delay Information</td></tr>
<tr><td></td><td>Material</td><td>Labor</td><td>Equipment</td><td>Management</td></tr>
<tr><td>C. Occurrences</td><td>2</td><td>3</td><td>1</td><td>1</td></tr>
<tr><td>D. Total Added Time</td><td>212</td><td>116</td><td>102</td><td>96</td></tr>
<tr><td>E. Probability of Occurrences[a]</td><td>0.167</td><td>0.25</td><td>0.0833</td><td>0.0833</td></tr>
<tr><td>F. Relative Severity[b]</td><td>0.805</td><td>0.294</td><td>0.774</td><td>0.729</td></tr>
<tr><td>G. Expected % Delay Time Per Production Cycle[c]</td><td>13.44</td><td>7.35</td><td>6.45</td><td>6.07</td></tr>
</table>

[a] Delay cycles/total number of cycles
[b] Mean added cycle time/mean overall cycle time
[c] Row E times row F times 100

Overall productivity = Ideal productivity (1 − Delays)
27.34 units/hr = 41.09 (1 − 0.1344 − 0.0735 − 0.0645 − 0.0607)

Index

Accounting 2, 37–34, 43, 46, 48, 54, 59, 72, 228
Adverse relationships 7, 32, 152
Advertising 74, 215
A/E 4, 6–7, 10–11, 26–27, 31, 57–58, 64–65, 84, 91, 132, 152–53, 164–65, 210, 220–21, 234, 240
AGC 42, 45–46, 269–83, 307–10
Agent 2–4, 28, 35, 38, 40, 58, 230, 245
Agreement 42
AIA 42, 45–46, 258–68, 307–10
American Institute of Architects. *See* AIA
Amortized 93, 113
Appraising 90
Appreciation 114, 118
Arbitrator 231, 311
Architect. *See* A/E
Associated General Contractors. *See* AGC
Authority 20, 38, 229–30, 245, 305

Backward pass. *See* Latest start time schedule
Balance sheet 37
Bank. *See* Lending institution
Bankruptcy 30
Bar chart 51, 55, 169–72, 235
Bidding 32, 90
Banking 13, 15, 52
Bonding company 1, 17
Bookkeeping 162
Brainstorming 155, 166, 234
Budgeting 49–50, 218

Capitol asset 100
Capital gain 87, 95, 101–102, 127
Cash flow 33, 88, 97, 99, 102, 109–12, 114, 120–21, 128–29, 207, 235, 237
Certified Public Accountant. *See* CPA
Change order 27–28, 31, 45, 56–57, 59–60, 208, 220, 321
Claims 56, 230, 310
Clean-up 62
CM
    advantages **23–41**
    contracts 43–66
    definition 2–12
    disadvantages **23–41**
    fee 24, 35–36, 39, 57, 68, 133, 304–305, 310
    firm 5, 7–9, 16, 18–21, 24–25, 151–152, 165–66, 209, 222, 231, 305
    process 1–22, 33, 208, 214, 241, 304
    proposal 77–84
    services 41–66, 169, 307–308
    skills 83–84, 98, 222
Committment letter 30
Communication 32–33, 152
Componentizing 48, 87, 97–100, 123–29, 314–15
Computer 146, 149, 161, 184, 206, 240
Conceptualizing 158
Confidence limit 147
Construction
    building 15
    heavy and highway 17, 40, 135
    private 8, 86

Construction (*continued*)
   loan 92
   public 8–9, 17, 71, 82, 86, 215
   specialty 33
Contract
   AGC 269–83, 307–10
   AIA 258–68, 307–310
   breach 3
   documents 46, 91, 135, 153, 229, 309
   drawings 86, 135, 151, 158, 164, 210, 234
   professional service 64
   guaranteed maximum 4–5
   specifications 151
Contractor
   failure 14
   general 2, 12, 14–19, 21, 24–25, 45, 54–55, 208, 213–15, 218–20, 229–230
   independent 2–4, 21
   prime 2, 18, 37, 45, 211, 244
   specialty 33, 37
   sub 15–16, 18, 24, 36, 98, 213, 218, 220, 229–30
   turnkey 2, 21
Control 12, 33, 44, 61, 133, 209, 212, 214, 216, 228–56
Correlation analysis 48
Cost
   accounting 229
   annual 92
   auditing 70, 72
   books 87–88, 90, 137
   building 90
   construction 92–95
   control system 228
   finance 89, 92–95, 241
   forms 235–40
   improvement 92
   maintenance and repair 87, 89, 95, 113–15
   of capital 49
   operating 95
   plus 30, 35
   replacement 95
   target 147–48

Cost of capital 87
CPA 84, 88
CPM 51, 55, 169–70, 172–207, 209, 223–226, 234–37, 320
Craftsmen 12
Critical activity 177
Critical path 183–84, 203, 224
Critical path method. *See* CPM

Debt service 86–87, 92, 113, 118
Delays 53, 60, 151, 177, 213, 254
Depreciation
   accelerated 97–98, 102, 126
   benefits 86, 90–98, 109–24
   componentized 48, 87, 97–100, 123–129
   excess 102, 119, 127
   methods 97–98, 109–24
   straight line 97, 120, 125
Design
   build 2, 5, 17, 210
   cost 91
   inefficiencies 11–12
   phase 10–12, 16, 49, 152–53, 208, 210
   ratios 140
   team 10
Designer. *See* A/E
Developer 71
Discounted cash flow 109–12
Documented 235–40
Dodge Service 73
Dummy activity 173
Duration 177–78, 212, 224
Dynamic programming 76

Earliest finish time 179, 181
Earliest start time 179, 181–82
Earliest start time schedule 179, 181, 185–186, 189–91, 202
Engineer. *See* A/E
Engineering News Record. See ENR
ENR 139
Environmental impact 49

Index

Estimate
    bill of material 154, 157–58
    CM 10–11, 25–26, 32, 132–50, 216, 241
    construction 56, 86
    detailed 142
    factor 133, 144–46
    feasibility 86–130, 132
    final 135
    function 136, 154
    parameter 43, 50–51, 139–44, 316
    preliminary 133–50
    range 133, 147–50
    technique 44
    trade 137–38
    unit cost 137, 139

Fast track. *See* Phased construction
Feasibility
    estimate 86–130, 314
    project 7, 43, 46
Federal Reserve System 93
FHA 93
Fiduciary 70
Financing 27, 48–49, 113, 237
Forward pass. *See* Earliest start time schedule
Four party process 17–18
Free float 181–83, 185
Free float prime 182–90
Frontending 240–41
Functional analysis 159–60

General conditions 141, 240
General Service Administration. *See* GSA
Governmental
    agency 27
GSA 16, 83
Guaranteed maximum 2, 4–5, 14, 64, 309

Hold harmless 32

Illinois Capital Development Board 73
Improvements 113
Information system 45
Inspection 25, 57–58, 65
Interest
    compound 104
    formuli 103–109
    simple 103–104
    tables 284–303
Internal Revenue Service. *See* IRS
Investment
    property 8, 63, 72
    tax credit 43, 48, 87
IRS 96, 98–99
ITC 99–101

Land 91–92
Latest finish time 182–84
Latest start time 182
Latest start time schedule 179, 182, 185–186, 189, 202
Leadership 57
Lending institution 1, 17, 241
Letter of transmittal 81–82
Letting 90
Liability 15, 42, 58, 64–65, 232, 305
Life cycle costs 86, 165
Liquidity 213
Long lead item 51
Loyalty 2–4

Maintenance and repair 87
Management
    practices 10–11
    services 38, 231
    skills 27, 40, 83–84
    techniques 234, 245
Market research 88
Marketing
    CM 68–85
    construction 5
    project 53, 62
    research 48
Markup 73

Material 29–31
Meetings 232–34
Method indicators 253
Method Productivity Delay Model. *See* MPDM
Miller Act 37
Minority set aside 45, 214, 216
Mortgage 119
Motion analysis 245–46
MPDM 245–56

Network model 172, 175
Nodes 179, 182

Observation. *See* Supervision
Occupational Safety and Health Act. *See* OSHA
Operating expense 115
Operating procedures 53, 56–57
Option clause 113
Ordinary gain 119, 127
Organization
  chart 82
  structure 17, 19–20, 23, 29, 215
OSHA 234
Overbilled 61–62, 240, 242
Overhead 18, 24, 26, 28, 215–16
Owner
  benefits 23
  needs 46
  project 8, 19, 28, 34, 35, 63, 164, 213
  purchased material 29, 31, 51–52, 222

Packaging 37, 43, 52, 147, 169–70, 208–26, 241, 316, 321
Paperwork 54, 59
Parameter estimating. *See* Estimating
Payback 116
Payment request 239
Performance bond 36, 213, 221
Permits 49, 53
Personnel management 234
Phased construction 1, 13, 21, 29, 43–44, 56, 133, 147, 169–70, 208–26

Planning 12, 28, 43, 45, 54–55, 169–207, 320
Preconstruction 7, 18, 53, 178
Privity of contract 18, 213, 215, 220
Probability 149
Process chart 212, 245
Process plants 144
Production
  cycle 246, 248–49
  equation 249–50
  unit 246
Productivity 11, 229, 245–55
Profit 20, 216
Progress
  billings 61
  payment 33, 134–35
Property management 43, 63, 73
Publicity 75
Purchase order 169

Qualifications 68–69
Queuing 245

Rate of return 109–12, 121–23, 128–29
Real estate 88, 90
Real estate investment trust. *See* REIT
Real property 63
Reasonable man concept 64
Recapture 101
Reimburseable 35
REIT 92
Renovation 72
Rental income 48
Reports
  daily 60
  progress 58–59
Request for proposal. *See* RFP
Resource
  allocation 174, 183–84, 187–205, 228, 321
  availability 193
  leveling 192, 201–206
  limited resource rule 192–201
  requirement 209
  scheduling method 193–201

Index

Responsibility accounting 217
Retainer 31, 33–34, 38, 61–62, 215
RFP 76

Safety 234
Scheduling 43, 45, 51, 55, 169–207, 211
Sixteen divisions 137–38
Subsidized loan 93
Substantial completion 33
Subsystem 155, 210
Superintendent 57
Supervision 35, 57–58, 65, 231, 240
Surety company. *See* Bonding company
Systems approach 7, 26

Tax
    analysis 47, 109–24
    benefits 95, 99–102
    considerations 86
    gain 119
    liability 115–24, 126
    savings 43, 46
    shelter 72

Testing materials 58–59
Three party process 16, 18
Time and material 30
Time lapse 249
Time scale 172, 174, 184–87
Time study 245–46
Time value of money 102–12, 121
Total float 183, 202–203, 206, 225–26
Turnkey construction 2, 5–6

Value engineering 43–44, 50, 151–68, 317–319

Wall Street Journal 74
Work improvement 254
Working capital 37
Work phases 170–207
Work sampling 345–46

Zoning 49